PHYSICS QUIZ

Edited By

R. Adamson

ANMOL PUBLICATIONS PVT. LTD.

NEW DELHI - 110 002 (INDIA)

ANMOL PUBLICATIONS PVT. LTD.

H.O.: 4374/4B, Ansari Road, Darya Ganj,
New Delhi-110 002 (India)
Ph.: 23278000, 23261597

B.O.: No. 1015, Ist Main Road, BSK IIIrd Stage
IIIrd Phase, IIIrd Block,
Bangalore - 560 085 (India)
Visit us at: www.anmolpublications.com

Physics Quiz

Reprint, 2007

PRINTED IN INDIA

Printed at Mehra Offset Press, Delhi.

Preface

Quiz series on Physics, Chemistry, Science, Mathematics, Botany and Zoology, has been written, compiled and edited by the experts of their respective fields. The purpose of the series is to enable students to assess their knowledge and understanding of the subject concerned. Each book includes more than 1000 questions. Every question included in this series has been extensively tested with school as well as college students. By means of these questions one can test not only factual recall but any other qualities that seem desirable, e.g., powers of deduction, ability to weigh evidence, apply known principles to new situations etc. Questions may be either factual of may involve a fair amount of reasoning before a decision can be made.

Editor

DIRECTIONS. *Each of the following questions is followed by four responses (answers). Select the correct response in each case. Show your answer by putting a cross against the response.*

Section I

Q. 1. The electrons behave as waves, because

(*a*) they can be diffracted by a crystal ☐

(*b*) they can produce ions in gases ☐

(*c*) they can be deflected by magnetic fields ☐

(*d*) they can be deflected by electric fields. ☐

Q. 2. In a hydrogen atom, the electron is moving in the xth orbit. The circumference of the orbit (x) and the de Broglie wavelength of the electrons are related as

(*a*) $x = nh$ ☐

(*b*) $x = n$ ☐

(*c*) $x = \frac{h}{n}$ ☐

(*d*) $x = nh^2$. ☐

Q. 3. Who gave the idea of matter wave ?

(*a*) de Broglie ☐

(*b*) Planck ☐

(*c*) Einstein ☐

(*d*) Huygen. ☐

Q. 4. The energy equivalent of 1 kg of matter is approximately

(*a*) 10^{-15} J ☐

(*b*) 1 J ☐

(*c*) 10^{-12} J ☐

(*d*) 10^{17} J. ☐

Q. 5. A football player kicked a football, football does not behave as a wave because the de Broglie wavelength corresponding to foot ball is

(*a*) infinite ☐
(*b*) much greater than the dimensions of the field but not infinite ☐
(*c*) of the same order as the dimensions of the field ☐
(*d*) much less than the dimensions of the field. ☐

Q. 6. The sun radiates energy at the rate of 3.6×10^{33} ergs/sec. This is equivalent to an annihilation of mass at the rate of

(*a*) 12×10^{10} kg/sec ☐
(*b*) 1.2×10^{13} g/sec ☐
(*c*) 4×10^{12} gm/sec ☐
(*d*) 8.5×10^{4} gm/sec. ☐

Q. 7. What is the energy associated with a photon of wavelength λ?

(*a*) $\frac{h}{\lambda}$ ☐
(*b*) $ch\lambda$ ☐
(*c*) $h\lambda$ ☐
(*d*) $\frac{hc}{\lambda}$. ☐

Q. 8. An electron microscope employs which one of the following principles ?

(*a*) Electrons have a wave nature ☐
(*b*) Electrons can be focused by an electric field ☐
(*c*) Electrons can be focused by magnetic field ☐
(*d*) All of the above. ☐

Q. 9. Wave nature of matter is not apparent in our day to day working because the wavelength of matter is

(*a*) directly proportional to energy ☐
(*b*) inversely proportional to energy ☐
(*c*) directly proportional to momentum ☐
(*d*) inversely proportional to momentum. ☐

Q. 10. According to theories of modern physics

(*a*) there is no such thing as particle ☐
(*b*) there is no such thing as wave ☐
(*c*) moving bodies behave in some respects as though they were waves and in others as though they were particles ☐

(*d*) all bodies, moving or stationary, behave in some respects as though they were waves and in others as though they were particles. ☐

Q. 11. When photoelectrons produced by light in films are fed into the loudspeaker, the loudness of sound depends upon

(*a*) the mass of the electrons ☐
(*b*) the number of electrons released ☐
(*c*) the wavelength of the original light ☐
(*d*) the charge of the electrons. ☐

Q. 11. Which of the following radiations will produce electrons with highest energy?

(*a*) Visible light ☐
(*b*) Ultraviolet light ☐
(*c*) X-rays ☐
(*d*) Gamma rays. ☐

Q. 13. The flow of heat from a hot to a cold body is an example of

(*a*) adiabatic process ☐
(*b*) irreversible process ☐
(*c*) reversible process ☐
(*d*) isothermal process. ☐

Q. 14. A system of molecules whose energies are distributed in the most probable way

(*a*) can perform an amount of mechanical work ☐
(*b*) can perform an amount of mechanical work equal to its total energy content ☐
(*c*) cannot perform any mechanical work ☐
(*d*) None of the above. ☐

Q. 15. A heat engine with 100% of efficiency would have to

(*a*) do no work ☐
(*b*) be at a uniform temperature ☐
(*c*) use no heat ☐
(*d*) discharge at 0° C. ☐

Q. 16. A temperature of 10^{-2} K can be reached by

(*a*) The Claude Heylandt method ☐
(*b*) Regenerative cooling ☐
(*c*) Adiabatic demagnetisation ☐
(*d*) Adiabatic desorption. ☐

Q. 17. The process in which no heat enters or leaves the system is termed as

(*a*) isochoric ☐
(*b*) isobaric ☐
(*c*) isothermal ☐
(*d*) adiabatic. ☐

Q. 18. The ratio between the energy dissipated in some process and the heat that appears as a result is the

(*a*) specific heat ☐
(*b*) triple point ☐
(*c*) kilocalorie ☐
(*d*) mechanical equivalent of heat. ☐

Q. 19. Whenever a system is made to complete a cyclic process, the work done during the complete cycle

(*a*) is zero ☐
(*b*) positive ☐
(*c*) is negative ☐
(*d*) depends upon the path followed. ☐

Q. 20. Two reversible engines *A* and *B* have their sources at 800 and 640 K. Their sinks are at 400K and 300 K respectively

(*a*) *A* is more efficient than *B* ☐
(*b*) *A* is less efficient than *B* ☐
(*c*) They are equally efficient ☐
(*d*) Their efficiencies depend only upon the type of fuels used. ☐

Q. 21. Streamlined objects move more easily through fluids than those of irregular shape. The reason is that

(*a*) turbulence is decreased ☐
(*b*) viscosity is increased ☐
(*c*) turbulence is increased ☐
(*d*) viscosity is decreased. ☐

Q. 22. Two pieces of wire of the same material have their length in the ratio of 1 : 2 and diameters in the ratio of 2 : 1. If they are stretched by the equal force, elongation will be in the ratio of

(*a*) 1 : 2 ☐
(*b*) 2 : 1 ☐

(c) 1 : 8 ☐
(d) 8 : 1. ☐

Q. 23. A shearing stress that acts on a body affects its
(a) width ☐
(b) length ☐
(c) volume ☐
(d) shape. ☐

Q. 24. The volume strain for a constant applied force increases directly with an increase in
(a) volume ☐
(b) compressibility ☐
(c) surface area ☐
(d) bulk modulus. ☐

Q. 25. Two wires, A and B, are made of the same material and are subjected to the same loads. The strain is greater for A when
(a) A has twice the diameter of B ☐
(b) A is twice as long as B ☐
(c) A has twice the length and half the diameter ☐
(d) A has twice the diameter and half the length. ☐

Q. 26. A shearing stress acting on a body changes its
(a) shape ☐
(b) length ☐
(c) volume ☐
(d) area. ☐

Q. 27. When impurities are added to metal, they
(a) will decrease the elasticity ☐
(b) will increase the elasticity ☐
(c) will not change the elasticity ☐
(d) will or will not change the elasticity. ☐

Q. 28. Four wires of the same material are stretched by the same load. The dimensions are given below. Which of them will elongate the most ?
(a) Length 200 cm, diameter 2 mm ☐
(b) Length 100 cm, diameter 1 mm ☐
(c) Length 300 cm, diameter 3 mm ☐
(d) Length 50 cm, diameter 0.5 mm. ☐

Q. 29. The height of water at two identical dams is the same, but dam A holds back a lake containing 2 cubic miles of water while dam B holds back a lake containing 1 cubic mile of water. The ratio of the total force exerted on dam A to that exerted on dam is B

(*a*) 2 ☐

(*b*) 1 ☐

(*c*) 1/2 ☐

(*d*) 4. ☐

Q. 30. Bernoulli's equation includes as a special case

(*a*) Hooke's law ☐

(*b*) Toricelli's theorem ☐

(*c*) Newton's third law of motion ☐

(*d*) Archimedes' principle. ☐

Q. 31. In a city water system the water will flow

(*a*) from the mains to the water outlet ☐

(*b*) only if the outlet is higher than the water in the water tower ☐

(*c*) faster from a first floor tap than from one on the third floor ☐

(*d*) only when the water tower has been completely emptied. ☐

Q. 32. The nib of a pen is split keeping in view the phenomenon of

(*a*) diffusion ☐

(*b*) osmosis ☐

(*c*) capillarity ☐

(*d*) cohesion. ☐

Q. 33. When there is no change in pressure at the beginning and end of a flow process, Bernoull's equation reduces to

(*a*) $P + 1/2\rho v^2 = \text{constant}$ ☐

(*b*) $P = \rho gh$ ☐

(*c*) $v = \sqrt{2gh}$ ☐

(*d*) None of these. ☐

Q. 34. For the Bernouli's theorem to be applicable, the flow of the liquid should be

(*a*) rotational and liquid should be incompressible ☐

(b) rotational and liquid should be compressible ☐
(c) irrotational and liquid should be incompressible ☐
(d) irrotational and liquid should be compressible. ☐

Q. 35. One of the important properties of electromagnetic waves is
(a) electric and magnetic fields are in the same phase ☐
(b) electric and magnetic fields are out of phase ☐
(c) electric and magnetic fields are some times in phase and some times out of phase. ☐
(d) All the above. ☐

Q. 36. Optical paths of monochromatic rays from a point object to its real point image through a converging lens are such that
(a) path for paraxial rays are shorter than the path for marginal rays ☐
(b) path for paraxial rays are longer than the path for marginal rays ☐
(c) they are equal for all rays through the lens ☐
(d) the equality or otherwise will depend on the focal length of the lens. ☐

Q. 37. Parallel rays of light are focused by a thin convex lens. A thin concave lens of the same focal length is then joined to the convex lens. The focal point will
(a) shift to infinity ☐
(b) shift towards the lens by a sma'! distance ☐
(c) shift away from the lens by a small distance ☐
(d) remain at its original position. ☐

Q. 38. The focal length of two lenses in contact is given by
(a) $F = f_1 + f_2$ ☐
(b) $F = f_1 - f_2$ ☐
(c) $F = f_1 . f_2$ ☐
(d) $\frac{1}{F} = \frac{1}{f_1} + \frac{1}{f_2}$. ☐

Q. 39. If D_1 and D_2 are the powers of two lenses placed in contact, then the power of the combination will be
(a) $D_1 + D_2$ ☐
(b) $D_1 - D_2$ ☐

(c) $\frac{D_1}{D_2}$ ☐

(d) $D_1 \times D_2$. ☐

Q. 40. A double convex air bubble in water acts as

(a) converging lens ☐
(b) diverging lens ☐
(c) plane slab ☐
(d) None of these. ☐

Q. 41. A biconvex lens of material of refractive index 1.5 has the radius of curvature of each side equal to 50 cm, the power of the lens will be

(a) 2 dioptre ☐
(b) 1 dioptre ☐
(c) 1.5 D ☐
(d) 0.5 D. ☐

Q. 42. What determines longitudinal chromatic aberration of a lens?

(a) Dispersive power only ☐
(b) Focal length only ☐
(c) Both dispersive power and focal length ☐
(d) None of these. ☐

Q. 43. The power of the lens in dioptres

(a) is its focal length in metres ☐
(b) is the reciprocal of its focal length in metres ☐
(c) is the reciprocal of length in metres ☐
(d) is the focal length in centimetres. ☐

Q. 44. A lens forms a sharp image on the screen on inserting a parallel sided slab of glass between the lens and the screen, it is found necessary to move the screen a distance '*d*' away from the lens in order to focus the image sharply. If the refractive index of glass relative to air is μ, then the thickness of glass slab is given by

(a) d/μ ☐
(b) μd ☐
(c) $\frac{\mu d}{\mu - 1}$ ☐

(*d*) $\frac{(\mu-1)d}{\mu}$.

Q. 45. The power of the lens which is the reciprocal of focal length is also given by the formulae in terms of refractive index and radius of curvature

(*a*) $(\mu-1)\left(\frac{1}{R_1}+\frac{1}{R_2}\right)$

(*b*) $\mu\left(\frac{1}{R_1}-\frac{1}{R_2}\right)$

(*c*) $\mu\left(\frac{R_1+R_2}{R_1}\right)$

(*d*) $(\mu-1)\left(\frac{1}{R_1}-\frac{1}{R_2}\right)$

Q. 46. Two convex lenses of focal length f used in combination become telescopic when the distance between them is

(*a*) f

(*b*) $4f$

(*c*) $2f$

(*d*) $f/2$.

Q. 47. A medium of dielectric constant '*K*' is introduced between plates of a parallel plate condenser. As a result its capacity

(*a*) increases '*k*' times

(*b*) decreases '*h*' times

(*c*) decreases $\frac{l}{k}$ times

(*d*) remains same.

Q. 48. Six equal resistances, each of 4 ohms are connected to form the Figure shown. The resistance between any two corners is

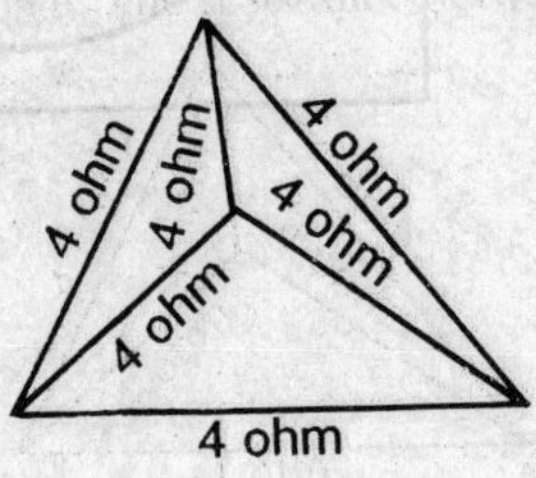

(*a*) 4 ohms
(*b*) 2 ohms
(*c*) 1 ohms
(*d*) $\frac{4}{6}$ ohms.

Q. 49. The magnetic flux density B at a distance r from a long straight wire carrying a steady flux density B at a distance r from a long straight wire carrying a straight conductor varies with r as shown in figure

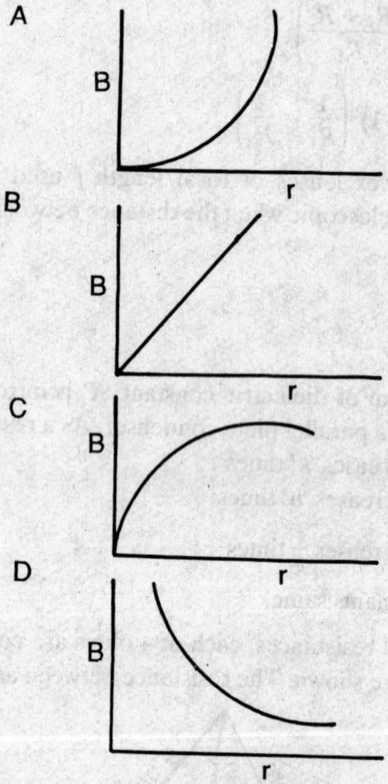

(*a*) *A*
(*b*) *B*
(*c*) *C*
(*d*) *D*.

Q. 50. An electron and a proton travel with equal speed in the same direction at 90° to a uniform magnetic field as this is switched on. They experience forces which are initially

(*a*) identical ☐
(*b*) equal but in opposite directions ☐
(*c*) in the same direction but differing by a factor of about 1840 ☐
(*d*) in opposite directions and differing by a factor of 1840. ☐

Q. 51. A coil of resistance R and inductance L is connected to a battery of E volt e.m.f. The final current in the coil

(*a*) $\frac{E}{R}$ ☐

(*b*) $\frac{E}{L}$ ☐

(*c*) $\frac{E}{R^2 + L^2}$ ☐

(*d*) $\frac{EL}{R^2 + L^2}$. ☐

Q. 52. If the instantaneous current in a circuit is given by $i = 2\cos(wt + \phi)$ amperes the r.m.s. value of current is given by

(*a*) 2 amperes ☐
(*b*) √2 amperes ☐
(*c*) 2√2 amperes ☐
(*d*) zero amperes. ☐

Q. 53. A current carrying conductor is placed in a uniform magnetic field experiences

(*a*) a mechanical push on it ☐
(*b*) an electrostatic imbalance ☐
(*c*) a strong force ☐
(*d*) All are wrong. ☐

Q. 54. The variation of potential with distance r from fixed point is as shown below

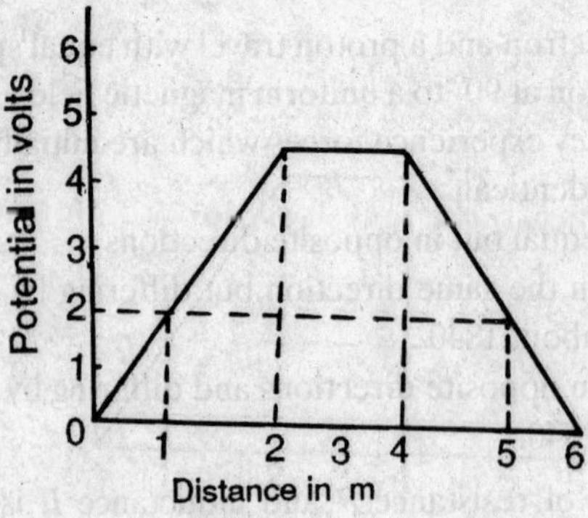

The electric field at $r = 5$ m is

(*a*) 2.0 volts /m

(*b*) -2.0 volts/m

(*c*) $\frac{2}{5}$ volts/m

(*d*) $-\frac{2}{5}$ volts/m.

Q. 55. Which of the following is a main feature of photoelectric effect ?

(*a*) Metals become conducting under the action of heat

(*b*) Electrons are ejected when light falls on some metals

(*c*) Electrons are emitted by heating of metals with ultraviolet rays

(*d*) Photographic images are formed under action of light rays.

Q. 56. Electron are emitted from a metal surface when light falling on it has a minimum

(*a*) energy

(*b*) wavelength

(*c*) velocity

(*d*) charge.

Q. 57. What is the mass of a photon of energy $h\nu$?

(*a*) $\frac{h\nu}{c^2}$

(b) $h\nu c^2$

(c) Infinite

(d) indefinite.

Q. 58. The energies of photo electrons in photoelectric effect

(a) changes with intensity of light

(b) changes with velocity of light falling as the metal surface

(c) changes with frequency of light

(d) None of the above is correct.

Q. 59. Which of the following functions is performed by a photo-cell ?

(a) It converts chemical energy into electrical energy

(b) It converts magnetic energy into electrical energy

(c) It converts light energy into electrical energy

(d) It converts electrical energy into light energy.

Q. 60. Photoelectric effect supports quantum nature of light because

(A) there is a minimum frequency of light below which no photoelectrons are emitted

(B) The maximum kinetic energy of photoelectrons depends only on the frequency of light and not on its intensity

(C) even when the metal surface is faintly illuminated, the photoelectrons leave the surface immediately

(D) electric charge of the photoelectrons is quantized

Which of the above is/are correct ?

(a) Only *A* is correct

(b) Both *A* and *B* are correct

(c) *A, B* and *C* are correct

(d) All the above are correct.

Q. 61. Einstein's photoelectric equation states that

$$E_k = h\nu - \phi$$

In this equation E_k refers to

(a) kinetic energy of all the emitted electrons

(b) mean kinetic energy of the emitted electrons

(c) maximum kinetic energy of the emitted electrons

(d) minimum kinetic energy of the emitted electrons.

Q. 62. The current in a photoelectric tube

(*a*) decreases with increase in frequency of the incident photons ☐

(*b*) increases with increase in frequency of the incident photons ☐

(*c*) decreases with increase in intensity of illumination ☐

(*d*) increases with increase in intensity of illumination. ☐

Q. 63. By which of the following can be verify the Einstein's photoelectric equation ?

(*a*) Milikan's experiment ☐

(*b*) Einstein's set up ☐

(*c*) Richardson's experiment ☐

(*d*) Thomson's experiment. ☐

Q. 64. In Compton scattering

(*a*) whole of the incident photon is absorbed ☐

(*b*) scattered photon has more energy than incident photon ☐

(*c*) whole of atom recoils ☐

(*d*) change in photon wavelength is independent of original wavelength of the photon. ☐

Q. 65. The maximum energy of emitted photoelectron is measured by

(*a*) the largest potential difference they can traverse ☐

(*b*) the current they produce ☐

(*c*) the potential difference they produce ☐

(*d*) the speed with which they emerge. ☐

Q. 66. The theory that light is emitted in discrete amounts of energy rather than in a continuous fashion is known as

(*a*) the photoelectric effect ☐

(*b*) the quantum theory ☐

(*c*) Huygen's principle ☐

(*d*) the electromagnetic theory. ☐

Q. 67. Which of the following colour phenomenon is *not* an interference phenomenon ?

(*a*) Oil spread on the surface of water exposed to extended sources of light exhibits brilliant colours ☐
(*b*) Soap bubbles in sunlight show colours ☐
(*c*) Sky appears blue at noon but red at evening and dawn ☐
(*d*) Metallic surface when heated show colours. ☐

Q. 68. In a thin film interference system
(*a*) the reflected system and transmitted system are same ☐
(*b*) the reflected system and transmitted system are opposite ☐
(*c*) some times they are same and some times they are opposite ☐
(*d*) None of the above is correct. ☐

Q. 69. The fact that light is transverse wave phenomenon is proved by the observation that light undergoes
(*a*) reflection ☐
(*b*) diffraction ☐
(*c*) polarisation ☐
(*d*) superimposition. ☐

Q. 70. When light passes close to the edge of any object, it is bent in its path and travels on in a new direction. This bending of light around corners is called
(*a*) refraction ☐
(*b*) polarisation ☐
(*c*) diffraction ☐
(*d*) interference. ☐

Q. 71. Fraunhoffer lines are evidence of
(*a*) the complete absence of certain elements in sun ☐
(*b*) the absence of certain elements in the sun's surface layers ☐
(*c*) the presence of certain elements in the sun's surface layers ☐
(*d*) the presence of certain elements in the interior of the sun. ☐

Q. 72. Air bubble in water shines because of
(*a*) reflection ☐

(*b*) refraction ☐
(*c*) diffraction ☐
(*d*) total internal reflection. ☐

Q. 73. In Michelson interferrometer experiment the scale reading for a pair of maximum indistinctness was formed as 0.6989 mm and 0.9884 mm. If the mean wavelength of two components of *D* lines be 5893 Å, the difference between the wavelenghts of components is
(*a*) 2 Å ☐
(*b*) 4 Å ☐
(*c*) 6 Å ☐
(*d*) 0.6 Å. ☐

Q. 74. Which is not an example of interference by division of wavefront ?
(*a*) Newton's rings ☐
(*b*) ,Fresnel Bi-prism ☐
(*c*) Young's slit ☐
(*d*) Lyold's mirror. ☐

Q. 75. A thin plate is introduced in the path of one of the beam of light in an interferrometer and it is found that 50 bands have crossed the line of observation. If the wavelength of light is 5896Å and $\mu = 1.4$ the thickness of the plate is
(*a*) 0.003685 cm ☐
(*b*) 0.00556 cm ☐
(*c*) 0.003885 cm ☐
(*d*) 0.00356 cm. ☐

Q. 76. The intensity of the dark band in interference pattern becomes zero when two waves
(*a*) of light are monochromatic ☐
(*b*) are of the same frequency ☐
(*c*) are of the same amplitude ☐
(*d*) travel in opposite direction. ☐

Q. 77. Interference of light waves results in alternate areas of minimum and maximum intensity means that
(*a*) energy at dark spots has been destroyed ☐
(*b*) energy at the bright spots has been created ☐

(*c*) there is transfer of energy from dark to bright region ☐
(*d*) average energy of medium is reduced. ☐

Q. 78. A Michelson interferrometer is illuminated by light of wavelength 6300 Å. The distance that mirror must move to cause 240 fringes to pass the interference points is
(*a*) 0.00656 cm ☐
(*b*) 0.00556 cm ☐
(*c*) 0.00756 cm ☐
(*d*) 0.00356 cm. ☐

Q. 79. Which of the following is *not* a statement of Kirchoff's law ?
(*a*) The algebraic sum of currents at any junction in a circuit, is zero ☐
(*b*) In any closed circuit, algebraic sum of e.m.f.'s is zero ☐
(*c*) Resistance of a conductor is equal to the ratio of voltage and current in that conductor ☐
(*d*) The current going into a junction is equal to the current leaving the junction. ☐

Q. 80. In the Wheatstone bridge shown in diagram what is the value of R_4 for a null, $R_1 = 5$ ohm, $R_2 = 10$ ohm, $R_3 = 15$ ohm
(*a*) 20 ohm ☐
(*b*) 30 ohm ☐
(*c*) 40 ohm ☐
(*d*) 250 ohm. ☐

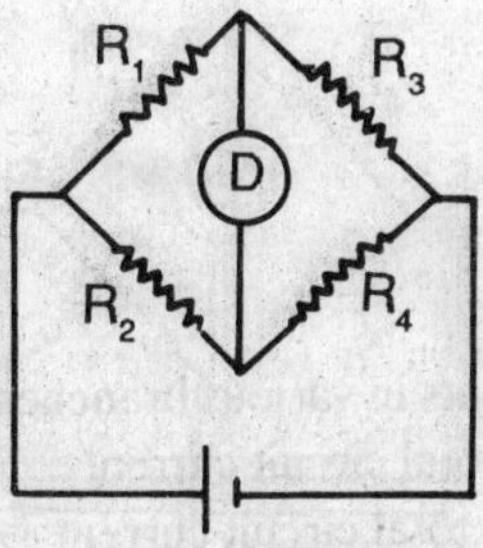

Q. 81. In the network shown below the effective resistance between *A* and *B* is

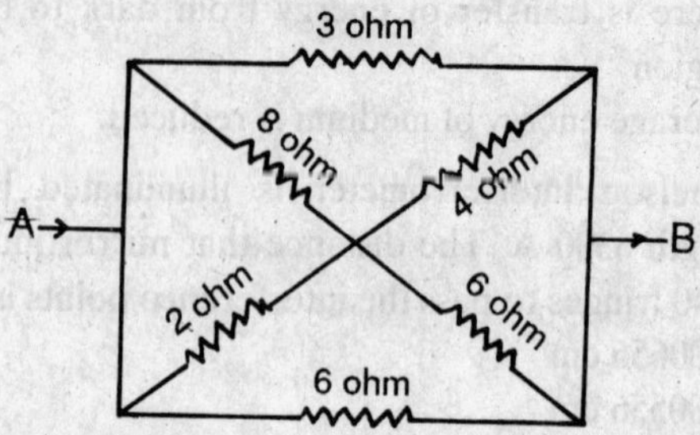

(*a*) $\frac{4}{3}$ ohm

(*b*) $\frac{3}{4}$ ohm

(*c*) $\frac{24}{17}$ ohm

(*d*) 29 ohm.

Q. 82. In the circuit shown below the current is

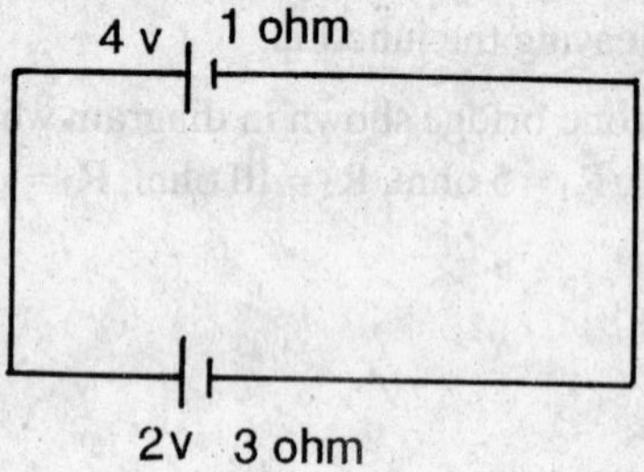

(*a*) 3 amp.

(*b*) 2 amp.

(*c*) 0.1 amp.

(*d*) 0.5 amp.

Q. 83. The sum of currents in various branches of the circuit

(*a*) equals the total circuit current

(*b*) exceeds the total circuit current

(*c*) less than the total circuit current

(*d*) None is correct.

Q. 84. The capacity of a parallel plate condenser filled with a dielectric medium of thickness '*d*' is given by

(*a*) $\dfrac{kA}{4\pi(t-d)}$ ☐

(*b*) $\dfrac{.A}{4\pi\left[t-d+\dfrac{d}{k}\right]}$ ☐

(*c*) Both the above ☐

(*d*) None of these. ☐

Q. 85. If a dielectric is placed between the plates of a capacitor

(*a*) its capacitance decreases ☐

(*b*) its capacitance increases ☐

(*c*) its capacitance is unaffected ☐

(*d*) None of these. ☐

Q. 86. The capacitance of a capacitor increases with a decrease in

(*a*) dielectric constant ☐

(*b*) permittivity ☐

(*c*) plate area ☐

(*d*) plate separation. ☐

Q. 87. Will a capacitor store more energy with a dielectric other than air ?

(*a*) Yes ☐

(*b*) No ☐

(*c*) Any of the ιwo ☐

(*d*) None of these. ☐

Q. 88. For polar substances dielectric constant is

(*a*) unity ☐

(*b*) infinite ☐

(*c*) zero ☐

(*d*) very large. ☐

Q. 89. A potentiometer can not be used for the measurement of

(*a*) current ☐

(*b*) voltage ☐

(*c*) internal resistance of cell ☐

(*d*) dielectric constant. ☐

Q. 90. The amount of charge which can be placed on a conductor does not depend on

(*a*) the dielectric strength of the surrounding medium ☐
(*b*) its capacitance ☐
(*c*) its potential ☐
(*d*) its size or shape. ☐

Q. 91. When a conductor situated in a dielectric is charged the energy resides

(*a*) only on the dielectric ☐
(*b*) only on medium surrounding charge ☐
(*c*) on the dielectric and medium both ☐
(*d*) only on the dielectric or medium surrounding charge. ☐

Q. 92. When we pass a radiation from a radioactive material through an electric field

(*a*) all three kinds of rays will be deflected ☐
(*b*) only the gamma rays are deflected ☐
(*c*) the α- and β-rays are deflected ☐
(*d*) only the α-rays are deflected. ☐

Q. 93. Radioactivity can be intensified by

(*a*) increasing the temperature of the element ☐
(*b*) increasing the electric field surrounding it ☐
(*c*) adiabatic demagnetisation ☐
(*d*) None of these. ☐

Q. 94. Which of the following causes the deflection of α-particles when they are passed through a thin foil ?

(*a*) Attraction to nucleus ☐
(*b*) Collison with nuclei ☐
(*c*) Interactions with electrons ☐
(*d*) Electrostatic repulsion by the nucleus. ☐

Q. 95. From which of the following, radium one of the radioactive elements, can be extracted ?

(*a*) Haematite ore ☐
(*b*) β-rays ☐

(c) Pitch blende ore ☐
(d) Magnetite ore. ☐

Q. 96. If an α-particle is heading straight for nucleus, when will it come to rest ?
(a) When its total energy is zero ☐
(b) When it strikes the nucleus ☐
(c) When it strikes the electron shells ☐
(d) When the kinetic energy lost by it is equal to electrical potential energy. ☐

Q. 97. The tunnel effect makes possible
(a) alpha decay ☐
(b) positive beta decay ☐
(c) negative beta decay ☐
(d) gamma decay. ☐

Q. 98. A freshly prepared radioactive source of half life 2 h emits radiation of intensity which is 64 times the permissible safe level. The minimum time after which it would be possible to work safely with this source is
(a) 6 h ☐
(b) 12 h ☐
(c) 24 h ☐
(d) 128 h. ☐

Q. 99. The β-ray emitted by a radioactive substance is
(a) an electron detached from an orbit outside the nucleus ☐
(b) an electron which was existing inside the nucleus ☐
(c) an electron emitted by the nucleus as a result of the decay of a neutron inside the nucleus ☐
(d) a pulse of electromagnetic wave. ☐

Q. 100. Particles which can be added to the nucleus of an atom without changing its chemical properties are called
(a) electrons ☐
(b) protons ☐
(c) neutrons ☐
(d) alpha particles. ☐

Section II

Q. 101. When a helium atom loses an electron it becomes
- (*a*) an alpha particle ☐
- (*b*) a proton ☐
- (*c*) a positive helium ion ☐
- (*d*) a negative helium ion. ☐

Q. 102. Emission of β-particle from a nucleus
- (*a*) changes its atomic number ☐
- (*b*) changes its mass number ☐
- (*c*) changes both the above ☐
- (*d*) None of these. ☐

Q. 103. Carbon dating method is possible because
- (*a*) C^{14} has higher atomic number than C^{12} ☐
- (*b*) C^{14} is quite stable ☐
- (*c*) C^{14} content of a dead body increases with time ☐
- (*d*) A dead body ceases to get C^{14} from the surroundings. ☐

Q. 104. Carbon content of a living object has at all times
- (*a*) some definite ratio of C^{14} to C^{12} atoms ☐
- (*b*) no definite ratio of C^{14} to C^{12} atoms ☐
- (*c*) no C^{14} at the time of birth of living object ☐
- (*d*) no C^{14} at the death of the living object. ☐

Q. 105. Who developed the radiocarbon dating of archeological samples ?
- (*a*) Otto Hahn ☐
- (*b*) W.F. Libby ☐
- (*c*) Enrico Fermi ☐
- (*d*) None of these. ☐

Q. 106. How is radiocarbon produced in the atmosphere ? It is produced by the bombardment of
- (*a*) oxygen by high energy neutrons ☐
- (*b*) oxygen by high energy protons ☐
- (*c*) nitrogen by high energy protons ☐
- (*d*) nitrogen by high energy neutrons. ☐

Q. 107. The capacity of a spherical conductor is numerically equal to its

(*a*) surface area ☐
(*b*) diameter ☐
(*c*) radius ☐
(*d*) volume. ☐

Q. 108. What do you call a resistance element side of a semi-conducting material and having a high negative coefficient of resistivity ?

(*a*) Thermistor ☐
(*b*) Insulator ☐
(*c*) Transistor ☐
(*d*) None of these. ☐

Q. 109. A capacitor is charged by using a battery which is then disconnected. A dielectric slap is then slipped between the plates which results in

(*a*) reduction of charges on the plates and increase of potential difference across the plates ☐
(*b*) increase in the potential difference across the plates, reduction in stored energy but no change in electric charge on the plates ☐
(*c*) decrease in the potential difference across the plates, reduction in stored energy but no change in the charge on the plates ☐
(*d*) None of these. ☐

Q. 110. The capacitance of a conductor in the presence of a dielectric medium

(*a*) increases with increase of dielectric constant ☐
(*b*) decreases with increase of dielectric constant ☐
(*c*) is proportional to the square root of dielectric constant ☐
(*d*) remains constant. ☐

Q. 111. The practical unit of capacity is

(*a*) Coulomb ☐
(*b*) Gauss ☐
(*c*) Henry ☐
(*d*) Farad. ☐

Q. 112. What happens while a plastic rod is being charged ?

(*a*) Electrons are removed or added ☐
(*b*) Electrons are given an additional negative charge ☐
(*c*) Protons are removed or added ☐
(*d*) Proton are given an additional positive charge. ☐

Q. 113. Condenser is used

(*a*) to produce charge ☐
(*b*) to change the direction of current ☐
(*c*) to collect the charge ☐
(*d*) as a good conductor of electricity. ☐

Q. 114. The force between two magnetic poles

(*a*) is directly proportional to the square of product of the strengths of the poles ☐
(*b*) is inversely proportional to the product of the strengths of the poles ☐
(*c*) is directly proportional to the square of the distance between them ☐
(*d*) is inversely proportional to the square of the distance between them. ☐

Q. 115. A field exerting a force of 1 dyne on a unit pole is known as

(*a*) magnetic moment ☐
(*b*) declination ☐
(*c*) magnetic force ☐
(*d*) oersted. ☐

Q. 116. A coulomb's torsion balance is used for

(*a*) verifying inverse square law ☐
(*b*) finding the mass of a magnet ☐
(*c*) comparing magnetic moments ☐
(*d*) measuring dip. ☐

Q. 117. A mariner's compass is used for

(*a*) verifying inverse square law ☐
(*b*) determining the direction ☐
(*c*) measuring dip ☐
(*d*) comparison of magnetic moments. ☐

Q. 118. The inverse square law in electricity can be verified by

(*a*) Whimshurt machine ☐

(*b*) gold leaf electroscope ☐
(*c*) coulomb's tension balance ☐
(*d*) Leyden jar. ☐

Q. 119. In a gaurd ring condenser the purpose of the gaurd ring is
(*a*) to raise the capacity ☐
(*b*) to lower the capacity ☐
(*c*) to avoid the variation of intensity at edges ☐
(*d*) to increase the effective area of the capacitor plates. ☐

Q. 120. Two condensers capacitance '*C*' each are connected in series. Their equivalent capacity is
(*a*) $C/2$ ☐
(*b*) C ☐
(*c*) $2C$ ☐
(*d*) C^2. ☐

Q. 121. Three identical capacitors of capacitance '*C*' each are connected in series and this combination is connected in parallel with one more identical capacitance. The capacitance of the whole combination is
(*a*) $3C$ ☐
(*b*) $3/C$ ☐
(*c*) $\frac{4C}{3}$ ☐
(*d*) $\frac{3C}{4}$. ☐

Q. 122. When there is resonance under forced oscillations,
(*a*) the frequency of oscillation is the undamped frequency under all conditions ☐
(*b*) when the damping is large, the oscillations are damped, and die down ☐
(*c*) when the damping is small, the amplitude of oscillations is large ☐
(*d*) the amplitude of oscillation is infinite at resonance for small damping. ☐

Q. 123. An example of forced oscillation is
(*a*) the pendulum of a clock ☐

(*b*) the windscreen-wiper

(*c*) a plucked sonometer wire

(*d*) a sonometer wire excited by a tuning fork.

Q. 124. For vibrating strings of the same length, the same stretching force and the same material, the relation between frequency *n* and diameter *D* is given by

(*a*) $n \propto \frac{1}{D}$

(*b*) $n \propto D$

(*c*) $n \propto \sqrt{D}$

(*d*) $n \propto \frac{1}{\sqrt{D}}$.

Q. 125. A sonometer wire vibrates in 3 loops. The frequency of vibration is given by

(*a*) $n = \frac{1}{2L} = \frac{3T}{m}$

(*b*) $n = \frac{1}{2L} = \frac{T}{3m}$

(*c*) $n = \frac{1}{2L} = \frac{T}{m}$

(*d*) $n = \frac{1}{6L} = \frac{T}{m}$.

Q. 126. If v_t and v_o are the velocities of sound in air at temperatures *T* and T_o degrees *A*

(*a*) $\frac{v_t}{v_0} = \frac{T}{T_0}$

(*b*) $\frac{v_0}{v_t} = \frac{T}{T_0}$

(*c*) $\frac{v_t}{v_0} = \frac{T_0}{T}$

(*d*) $\frac{v_t}{v_0} = \frac{T}{T_0}$.

Q. 127. The velocity of a longitudinal wave in a solid is given by

(*a*) $v = \frac{K}{\rho}$, where *K* is the bulk modulus

(b) $v = \sqrt{\dfrac{Y}{\rho}}$, where Y is the Young's modulus ☐

(c) $v = \sqrt{\dfrac{K}{Y}}$ ☐

(d) $v = \sqrt{\dfrac{Y}{K}}$. ☐

Q. 128. Two tuning forks A and B give 4 beats per second. On loading A slightly we get 3 beats in 5 seconds. The frequency of B is 256. What is the frequency of A after loading ?

(a) 260 Hz ☐
(b) 252 Hz ☐
(c) 256.6 Hz ☐
(d) 255.4 Hz. ☐

Q. 129. A rod clamped at the midpoint can be excited to produce longitudinal stationary waves. The frequency is given by $n = \dfrac{s}{2L}\sqrt{\dfrac{Y}{P}}$. In this expression

(a) s is the number of loops ☐
(b) s is the number of antinodes ☐
(c) s is the number of nodes ☐
(d) s is the number of nodes and antinodes. ☐

Q. 130. A tuning fork is a case of

(a) a rod with longitudinal vibrations ☐
(b) two fixed free rods with transverse vibrations ☐
(c) vibrations of a rod rich in overtones ☐
(d) frequency being independent of temperature. ☐

Q. 131. A sensitive bolometer is based on

(a) variation of resistance with temperature ☐
(b) the differential air thermometer ☐
(c) thermo e.m.f. ☐
(d) photochemical effect. ☐

Q. 132. The radiometer is an instrument for

(a) measuring the wavelength of radio waves ☐
(b) detecting radio activity ☐
(c) measuring the dosage of X-radiation ☐
(d) measuring infrared radiations. ☐

Q. 133. A perfectly black body

(*a*) is a perfect absorber of all radiations ☐
(*b*) is a perfect reflector of radiations ☐
(*c*) is a perfect reflector of visible light radiations ☐
(*d*) is a brownish object. ☐

Q. 134. The practical black body is

(*a*) a circular silver disc painted black ☐
(*b*) a glass bulb coated with lamp black ☐
(*c*) a polished copper sphere with a small black part ☐
(*d*) a hollow sphere with the inside coated black and with a small hole at the surface. ☐

Q. 135. According to the Prevost's theory of exchanges

(*a*) a hot body only radiates ☐
(*b*) a cold body only absorbs radiation ☐
(*c*) the radiation of a body depends on the nature and proximity of the surroundings ☐
(*d*) the amount of energy emitted depends on the nature and temperature of the body. ☐

Q. 136. According to Stefan's Law of radiation, the energy E emitted by unit area of a hot surface is given by $E = \alpha T^4$

(*a*) true ☐
(*b*) false ☐
(*c*) uncertain ☐
(*d*) None of these. ☐

Q. 137. Which of the following is *not* a ferromagnetic material ?

(*a*) Iron ☐
(*b*) Steel ☐
(*c*) Copper ☐
(*d*) Cobalt. ☐

Q. 138. Reverberation is the

(*a*) presence of large number of overtones ☐
(*b*) presence of harsh and discordant notes ☐
(*c*) the persistance of audible sound after the source has stopped ☐
(*d*) presence of ultrasonic vibrations. ☐

Q. 139. Good acoustics implies

(*a*) obtaining as much reverberation as possible ☐
(*b*) making the reverberation as small as possible ☐
(*c*) obtaining the optimum amount of reverberation ☐
(*d*) eliminating reverberation. ☐

Q. 140. Point out the wrong statement: Sabine's Law is

$$T = \frac{0.05\, V}{\Sigma\, a\, S_1}$$

(*a*) T is the period of reverberation ☐
(*b*) V is the volume of the hall in m^3 ☐
(*c*) a is the absorption coefficient ☐
(*d*) S_1 is the area of the surface having absorption coefft α. ☐

Q. 141. In the sound box of the early gramophone

(*a*) the mechanical vibrations were converted into varying currents ☐
(*b*) varying currents were amplified ☐
(*c*) the vibrations of the diaphragm produced sound waves in the horn ☐
(*d*) the needle was a permanent fixture. ☐

Q. 142. The reproduction of sound from the sound track of a film is done with

(*a*) a microphone ☐
(*b*) a pick up ☐
(*c*) a photoelectric cell ☐
(*d*) a string galvanometer. ☐

Q. 143. *Match the following*

I. Carbon microphone	(A) Ribbon
II. Dynamic microphone	(B) Charging
III. Condenser microphone	(C) Permanent magnet
IV. Velocity microphone	(D) Telephone

(*a*) *I-A, II-B, III-C, IV-D* ☐
(*b*) *I-B, II-C, III-D, IV-D* ☐
(*c*) *I-C, II-D, III-A, IV-B* ☐
(*d*) *I-D, II-C, III-B, IV-A.* ☐

Q. 144. The substance with which a recording tape is coated is

(*a*) soft iron ☐
(*b*) ferrite or iron oxide ☐
(*c*) pure magnetic steel ☐
(*d*) silver nitrate. ☐

Q. 145. The fundamental tone in a closed pipe has the frequency of 200 Hz. Then the second overtone

(*a*) is not present ☐
(*b*) has a frequency 1000 Hz ☐
(*c*) has a frequency 400 Hz ☐
(*d*) has a frequency 600 Hz. ☐

Q. 146. A closed pipe vibrates in its first overtone. Then the length of the pipe is given by

(*a*) $L = \frac{\lambda}{2}$ ☐
(*b*) $L = \frac{\lambda}{4}$ ☐
(*c*) $L = \frac{3}{4}\lambda$ ☐
(*d*) $L = \frac{5}{4}\lambda.$ ☐

Q. 147. A man standing on a railway track notices the change of pitch of the note due to the whistle of an engine as it passes him. If the frequency of the whistle by 256 Hz. When can be the frequency of the note heard by him when the train is approaching him ?

(*a*) 269.5 Hz ☐
(*b*) 243.8 Hz ☐
(*c*) 243.2 Hz ☐
(*d*) 250.2 Hz. ☐

Q. 148. A police van is sounding its siren at 512 Hz, while overtaking a speeding motorist. If the speedometer of the police van indicates 72 km/h and that of the other car 54 km/h, what is the apparent frequency of the siren to the motorist ? Velocity of sound in air = 340 m/s at room temperature

(*a*) 520 Hz ☐
(*b*) 568 Hz ☐

(c) 462 Hz ☐
(d) 504 Hz. ☐

Q. 149. Speeds of bodies exceeding the speed of sound are called
(a) supersonic ☐
(b) ultrasonic ☐
(c) infrasonic ☐
(d) superfast. ☐

Q. 150. Ultrasonic waves can be generated by a
(a) piece of glass ☐
(b) quartz crystal oscillator ☐
(c) car horn ☐
(d) a radio loudspeaker. ☐

Q. 151. When mechanical pressure is applied to the opposite faces of a suitably cut quartz crystal, a p.d. is produced between the faces
(a) true ☐
(b) false ☐
(c) uncertain ☐
(d) None of these. ☐

Q. 152. A bar or tube of a ferromagnetic material changes in length when it is placed in a strong magnetic field perpendicular to its lengths
(a) true ☐
(b) false ☐
(c) uncertain ☐
(d) None of these. ☐

Q. 153. A point charge is placed at the centre of a spherical Gausian surface. In which of the following cases will the value of ϕE, the flux, change
(a) if the surface is replaced by a cube of the same value ☐
(b) if the sphere is replaced by a cube of one-tenth the value ☐
(c) if the charge is moved off-centre in the original sphere ☐
(d) if the charge is moved just outside the original sphere. ☐

Q. 154. The north pole of a bar magnet having a pole strength of 600 unit poles is placed 10 cm from the south pole of another bar magnet whose pole strength is 400 unit poles. The force of attraction between the poles equals

(*a*) 150 dynes ☐

(*b*) 66g ☐

(*c*) 24×10^2 dynes ☐

(*d*) 24×10^2 g. ☐

Q. 155. When the force of attraction between two electrostatic charges is expressed in newtons and the distance between charges in metres the unit of charge is expressed in

(*a*) stat coulombs ☐

(*b*) coulombs ☐

(*c*) micro-coulombs ☐

(*d*) electrostatic units. ☐

Q. 156. We known the units for the constant of proportionality used in coulombs law are

(*a*) Nm^2 ☐

(*b*) Nm^2C^{-2} ☐

(*c*) N ☐

(*d*) None of these. ☐

Q. 157. An electric current exists in a long wire. At a distance of 2 cm from the wire the magnetic field is 0.2 gauss. The field at 4 cm from the wire is

(*a*) $\frac{1}{20}$ gauss ☐

(*b*) 0.4 gauss ☐

(*c*) 0.1 gauss ☐

(*d*) 0.05 gauss. ☐

Q. 158. A coil of are turn and 34 cm radius carries a current of 8 amp. The magnetic field at the centre of the coil in oerstead is

(*a*) $\frac{\pi \times 1 \times 8}{10 \times 34}$ ☐

(*b*) $\frac{8\pi}{34}$ ☐

(*c*) $\frac{2\pi \times 8}{34}$ ☐

(*d*) $\frac{2\pi \times 1 \times 8}{10 \times 34}$. ☐

Q. 159. The field inside a hollow spherical conductor is

(*a*) constantly zero ☐
(*b*) constant but not necessarily zero ☐
(*c*) a function of charge on sphere ☐
(*d*) a function of distance from the centre. ☐

Q. 160. All bulbs in Figure below are identical. Which bulb(s) light(s) most brightly

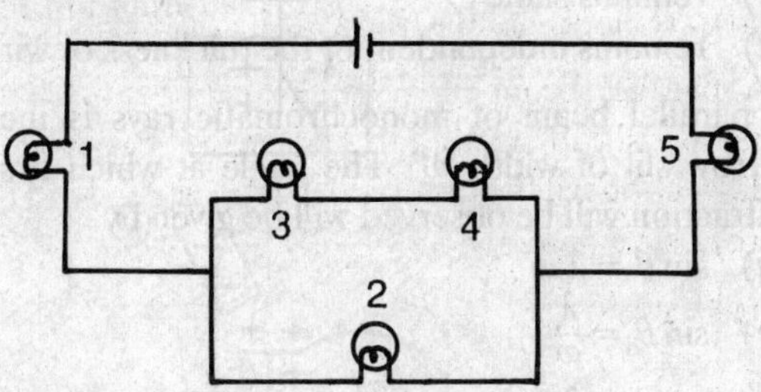

(*a*) 1 only ☐
(*b*) 2 only ☐
(*c*) 3 and 4 ☐
(*d*) 1 and 5. ☐

Q. 161. Rising or setting sun appears to be reddish due to the fact that

(*a*) the sun is colder at sun rise and sun set ☐
(*b*) diffraction sends red rays to the earth at the sun rise and sun set ☐
(*c*) refraction of light rays ☐
(*d*) scattering of light rays. ☐

Q. 162. In a diffraction pattern, the width of any fringe is

(*a*) directly proportional to slit width ☐
(*b*) inversely proportional to slit width ☐
(*c*) has no dependance on slit width ☐
(*d*) All the above. ☐

Q. 163. When the moon is near the horizon it appears bigger. This is due to

(*a*) atmospheric refraction ☐
(*b*) scattering of light ☐
(*c*) diffraction ☐
(*d*) optical illusion. ☐

Q. 164. While studying diffraction pattern with a wire we find the fringe width .... on increasing the diameter of the wire

(*a*) also increases ☐
(*b*) decreases ☐
(*c*) remains same ☐
(*d*) remains independent of the thickness of wire. ☐

Q. 165. A parallel beam of monochromatic rays is incident on a narrow slit of width 'ω'. The angle at which first minimum diffraction will be observed will be given by

(*a*) $\sin\theta = \lambda.\omega$ ☐
(*b*) $\sin\theta = \frac{\lambda}{\omega}$ ☐
(*c*) $\sin\theta = \frac{\omega}{\lambda}$ ☐
(*d*) $\sin\theta = \frac{2\omega}{\lambda}$. ☐

Q. 166. The radii of half period zones in the phenomenon of diffraction depends on the number of zones as

(*a*) inversely proportional to $\frac{1}{\sqrt{n}}$ ☐
(*b*) directly proportional to $\sqrt{n}$ ☐
(*c*) inversely proportional to $\frac{1}{(n)^{3/2}}$ ☐
(*d*) directly proportional to $\frac{1}{(n)^{-3/2}}$. ☐

Q. 167. How many orders will be visible if the wavelength of incident radiation be 5000 Å and number of lines on grating be 7620 to an inch ?

(*a*) Six orders ☐
(*b*) Four orders ☐
(*c*) Three orders ☐
(*d*) Five orders. ☐

Q. 168. Diffraction is the property according to which light waves

(*a*) change their direction on entering a different medium

(*b*) produce chemical effects

(*c*) bend round the corners

(*d*) None of these.

Q. 169. A Rowland grating has 15,000 lines to the inch. Calculate the angular separation of two mercury lines 5770 Å and 5461 Å in the spectrum of first order

(*a*) 2.83×10^{-2} radians

(*b*) 4.83×10^{-2} radians

(*c*) 1.83×10^{-2} radians

(*d*) 3.83×10^{-2} radians.

Q. 170. In a diffraction pattern of a single slit the width of the central fringe is

(*a*) twice as other fringes

(*b*) of the same size as other fringes

(*c*) of negligible size

(*d*) All the above.

Q. 171. In order to supress second order spectrum in the case of a grating

(*a*) the width of ruled portion the width of unruled portion

(*b*) the width of ruled portion the width of unruled portion

(*c*) the width of ruled portion = the width of unruled portion

(*d*) All the above are correct.

Q. 172. The dispersive power of a grating is directly proportional to

(*a*) wavelength

(*b*) separation of lines

(*c*) frequency of light used

(*d*) is independent of light used.

Q. 173. The spectrum obtained with a grating is called

(*a*) anomalous spectrum

(*b*) normal spectrum

(c) impure spectrum ☐
(d) grating spectrum. ☐

Q. 174. In a Young's double slit experiment the fringe width was 0.20 mm when observed using green light. If red light replaces green light, the fringe width becomes

(a) 0.25 mm ☐
(b) 0.16 mm ☐
(c) 0.13 mm ☐
(d) 0.10 mm. ☐

Q. 175. Dispersive power of a grating increases with

(a) order of spectrum ☐
(b) number of lines per centimetre ☐
(c) order and number of lines per centimetre ☐
(d) total number of lines on the grating. ☐

Q. 176. Fresnel diffraction is

(a) a near field phenomenon ☐
(b) a far field phenomenon ☐
(c) Both the above ☐
(d) None of these. ☐

Q. 177. How is radiocarbon dating done ?

(a) By estimating the amount of radium present in sample ☐
(b) By estimating the ratio of amounts C^{14} and C^{12} still present in the sample ☐
(c) By estimating the radiocarbon still present in the sample ☐
(d) By estimating the ordinary carbon still present in the sample. ☐

Q. 178. Which of the following propagate at the same speed as velocity of light ?

(a) Heat waves ☐
(b) Sound waves ☐
(c) Shock waves ☐
(d) β-rays. ☐

Q. 179. The phenomenon of radioactivity is associated with

(a) decay of nucleus ☐

(*b*) decay of atoms ☐
(*c*) fusion of nuclei ☐
(*d*) emission of electrons. ☐

Q. 180. Carbon 14 is found in living organisms in the amount of 100 atoms of carbon 14 for every 10^{20} atoms of carbon 12. Carbon 14 emits beta particles and has a half-life of about 6,000 years. The count rate from carbon 14 in a fossil indicates that the amount of carbon 14 atoms in the fossil has decreased to about 10 atoms of carbon 14 for every 10^{20} atoms of carbon 12. What is the best estimate of the age of the fossil ?

(*a*) 600 years ☐
(*b*) 20,000 years ☐
(*c*) 30,000 years ☐
(*d*) 54,000 years. ☐

Q. 181. The process of fission is responsible for the release of energy in

(*a*) the hydrogen bomb ☐
(*b*) the atom bomb ☐
(*c*) the sun ☐
(*d*) All the above. ☐

Q. 182. What are the values of *A* and *Z* in the nuclear reaction ?

$_{90}Tu^{234} \rightarrow Z_ZPa^A + {}_{-1}e^0 + v_0^0$

(*a*) $A = 234, Z = 91$ ☐
(*b*) $A = 233, Z = 90$ ☐
(*c*) $A = 232, Z = 92$ ☐
(*d*) $A = 235, Z = 80$. ☐

Q. 183. Who designed the atomic reactor ?

(*a*) Wilson ☐
(*b*) Rutherford ☐
(*c*) Fermi ☐
(*d*) Teller. ☐

Q. 184. In which of the following does the nuclear force exist ?

(*a*) Neutron-neutron ☐
(*b*) Proton-proton ☐
(*c*) Neutron-proton ☐
(*d*) All are true. ☐

Q. 185. For fission to occur, neutrons must possess

(*a*) very low energy ☐
(*b*) thermal energy ☐
(*c*) very high energy ☐
(*d*) None of these. ☐

Q. 186. During a nuclear fusion reaction

(*a*) a heavy nucleus breaks into two fragments by itself ☐
(*b*) a light nucleus bombarded by thermal neutrons breaks up ☐
(*c*) a heavy nucleus bombarded by thermal neutrons breaks up ☐
(*d*) two light nuclei combine to give a heavier nucleus and possibly other products. ☐

Q. 187. Transuranic elements are those

(*a*) which have atomic number as 92 ☐
(*b*) which have atomic number more than 92 ☐
(*c*) which have atomic number less than 92 ☐
(*d*) None of the above. ☐

Q. 188. If the number of turns in a coil is tripled the magnetic flux linked with it

(*a*) remains unchanged ☐
(*b*) tripled ☐
(*c*) becomes one third ☐
(*d*) None of these. ☐

Q. 189. Which of the following has the maximum permeability ?

(*a*) Paramagnetic substance ☐
(*b*) Ferromagnetic substance ☐
(*c*) Diamagnetic substance ☐
(*d*) None of these. ☐

Q. 190. The permeability of diamagnetic material is

(*a*) greater than unity ☐
(*b*) less than unity ☐
(*c*) equal to unity ☐
(*d*) zero. ☐

Q. 191. For a paramagnetic substance the correct relationship between (X) magnetic susceptibility and (T) absolute temperature is

(*a*) $X \propto T$ ☐

(*b*) $X \propto$ exp. (constant α T) ☐

(*c*) $X \propto \frac{1}{T}$ ☐

(*d*) X = constant. ☐

Q. 192. When two suspended objects are seen to attract each other electrically

(*a*) they are both charged ☐

(*b*) are must be charged ☐

(*c*) either (*a*) or (*b*) is charged ☐

(*d*) neither (*a*) nor (*b*) is true. ☐

Q. 193. Select the one that is paramagnetic ?

(*a*) Bismuth ☐

(*b*) Chromium ☐

(*c*) Water ☐

(*d*) Mercury. ☐

Q. 194. Dipole moment of a current loop is given by

(*a*) ncA ☐

(*b*) nqA ☐

(*c*) $\frac{n}{q}.A$ ☐

(*d*) All the above. ☐

Q. 195. Thermopile is used for

(*a*) converting atomic energy into heat energy ☐

(*b*) measuring the radiant energy ☐

(*c*) storing the heat energy ☐

(*d*) measuring current. ☐

Q. 196. The phase angle between the voltage and current in A.C. circuit through a resistor is

(*a*) 90° ☐

(*b*) 180° ☐

(*c*) zero ☐

(*d*) 60°. ☐

Q. 197. Gauss's law is given by

(*a*) $\varepsilon_0 \int \vec{E}.\vec{dS} = 0$ ☐

(*b*) $\varepsilon_0 \int \vec{E}.\vec{dS} = q$ ☐

(*c*) $\varepsilon_0^2 \int \vec{E}.\vec{dS} = q$ ☐

(*d*) $\int \vec{E}.\vec{dS} = \frac{q^2}{\varepsilon_0}$. ☐

Q. 198. Conversion of heat into electrical energy may be achieved by using

(*a*) hydrometer ☐
(*b*) thermocouple ☐
(*c*) voltmeter ☐
(*d*) photoelectric tube. ☐

Q. 199. Thermocouple is an arrangement of two different metals

(*a*) to convert heat energy into electric energy ☐
(*b*) to produce more heat ☐
(*c*) to convert heat energy into chemical energy ☐
(*d*) to convert electrical energy to heat energy. ☐

Q. 200. The magnitude of induced e.m.f. during electromagnetic induction is controlled by

(*a*) magnetic flux ☐
(*b*) electric flux ☐
(*c*) electric field ☐
(*d*) magnetic field. ☐

Section III

Q. 201. When a diamagnetic substance is inserted in current carrying coil, the magnetic field is

(*a*) decreased ☐
(*b*) unchanged ☐
(*c*) increased ☐
(*d*) increased or decreased, depending upon the relative volume of the substances. ☐

Q. 202. During severe winter in the temperate zones of the world, the superficial parts of the lakes are frozen, leaving water below. The freezing at the bottom is prevented because

(*a*) the conductivity of ice is low ☐

(*b*) the water has large latent heat of fusion ☐

(*c*) the water has large specific heat ☐

(*d*) the temperature of the earth at the bottom of the lake is high . ☐

Q. 203. Which of the following geometries will result in the largest convection coefficient ?

(*a*) Vertical plate ☐

(*b*) Horizontal plate, facing upward ☐

(*c*) Diagonal plate ☐

(*d*) Horizontal plate, facing downward. ☐

Q. 204. If a black body at absolute temperature T is surrounded by another black body at absolute temperature T_o, the amount of energy E lost per second per square meter by the body at higher temperature is proportional to

(*a*) $(T - T_0)^{1/4}$ ☐

(*b*) $(T^4 - T_O^4)$ ☐

(*c*) $(T - T_O)^4$ ☐

(*d*) None of above. ☐

Q. 205. If the amount of radiant energy received per unit area per second from the sun is measured at the Earth, Mars and Jupiter, it will be

(*a*) in the decreasing order Mars, Earth, Jupiter ☐

(*b*) in the increasing order Jupiter, Mars, Earth ☐

(*c*) the same at all the three ☐

(*d*) None of these. ☐

Q. 206. A body which absorbs all the radiation falling on it without reflecting or transmitting any of it, is called a

(*a*) luminous body ☐

(*b*) white body ☐

(*c*) perfectly black body ☐

(*d*) None of above. ☐

Q. 207. Heat produced by friction is transferred by

(*a*) conduction ☐

(b) radiation ☐
(c) convection ☐
(d) all of these. ☐

Q. 208. The law "wave length corresponding to a maximum energy is inversely proportional to the absolute temperature of the black body" is known as

(a) Wien's law ☐
(b) Kirchoff's law ☐
(c) Stefan's law ☐
(d) Newtons law of cooling. ☐

Q. 209. There are two thermometers based on different thermometric properties of two materials. They show identical readings because

(a) each property changes uniformly with temperature ☐
(b) the relation between the property and temperature is identical in the two cases ☐
(c) the property of one increases with temperature and the property of the other decreases at a uniform rate ☐
(d) the two thermometers have been calibrated with reference to a common standard. ☐

Q. 210. The standard fixed point for calibrating a thermometer is

(a) the boiling point of water ☐
(b) the boiling point of ice ☐
(c) the temperature of steam ☐
(d) the triple point of water. ☐

Q. 211. The gas thermometer is taken as the primary standard because

(a) the thermometers are easily reproducible ☐
(b) readings can be accurately taken ☐
(c) no corrections are necessary ☐
(d) it reproduces the thermodynamic scale. ☐

Q. 212. In a four stroke cycle, the crank makes

(a) half a revolution ☐

(*b*) two complete revolution

(*c*) one complete revolution

(*d*) four complete revolutions.

Q. 213. The Carnot cycle is a

(*a*) reversible cyclic process with two isotherms and two adiabats

(*b*) constant pressure cycle

(*c*) constant-volume cycle

(*d*) a reversible two-stroke cycle.

Q. 214. The process by which nuclei emit α, β- and γ-rays in order to attain stability is called

(*a*) radioactive disintegration

(*b*) β-decay

(*c*) radioactivity

(*d*) radioactive transmutation.

Q. 215. Atomic explosion occurs due to

(*a*) uncontrolled chain reaction in fission

(*b*) uncontrolled chain reaction in fusion

(*c*) controlled chain reaction in fission

(*d*) controlled chain reaction in fusion.

Q. 216. The first phase of a nuclear reaction involves the formation of

(*a*) an α-particle

(*b*) a compound nucleus

(*c*) an isotope

(*d*) a meson.

Q. 217. The process by which a heavy nucleus splits up into two lighter nuclei is known as

(*a*) fission

(*b*) fusion

(*c*) α-decay

(*d*) a chain reaction.

Q. 218. It is possible to understand nuclear fission on the basis of

(*a*) independent particle model of the nucleus

(*b*) liquid drop model of the nucleus

(*c*) meson theory of nucleus

(*d*) proton-proton type.

Q. 219. The phenomenon of nuclear fusion is used in the construction of

(*a*) incendiary bombs ☐
(*b*) atom bombs ☐
(*c*) hydrogen bombs ☐
(*d*) I.C.B.M. ☐

Q. 220. The control rods used in nuclear reactors are made of

(*a*) stainless steel ☐
(*b*) graphite ☐
(*c*) cadmium ☐
(*d*) plutonium. ☐

Q. 221. Which atomic particle is used to cause fission in an atomic reactor ?

(*a*) Alpha particle ☐
(*b*) Beta particle ☐
(*c*) Deuteron ☐
(*d*) Neutron. ☐

Q. 222. (1) Electrons are the decay product of neutrons
(2) In β-decay electrons come out from the nucleus
(3) There is no electron inside the nucleus
State if

(*a*) only (2) is correct ☐
(*b*) only (3) is correct ☐
(*c*) Both (2) and (3) are correct ☐
(*d*) All the above are correct. ☐

Q. 223. The de Broglie wavelength of a particle of mass m and energy E is

(*a*) $\sqrt{2hmE}$ ☐
(*b*) $\frac{h}{\sqrt{2mE}}$ ☐
(*c*) $\frac{1}{\sqrt{2hmE}}$ ☐
(*d*) $\frac{E}{\sqrt{2hm}}$. ☐

Q. 224. Release of fission energy results because of destruction of ..... in the nucleus ?

(a) Gravitational forces □
(b) Electrostatic forces □
(c) Mass □
(d) None of these. □

Q. 225. Moderators in a fission process slow down the fast neutrons and make it easy to produce fission in

(a) U^{235} □
(b) Th^{223} □
(c) natural uranium □
(d) U^{239}. □

Q. 226. A capacitor and an iron inductor are connected in series with a constant voltage, variable-frequency supply. The frequency is adjusted to give maximum current. How is the current and its frequency now affected by the removal of the iron core of the inductor ?

	Current	*Frequency*	
(a)	Increases	No change	□
(b)	Decreases	Increases	□
(c)	Decreases	No change	□
(d)	Decreases	Decreases.	□

Q. 227. An A.C. voltage is applied across the terminals *XY* in the circuit in Figure below

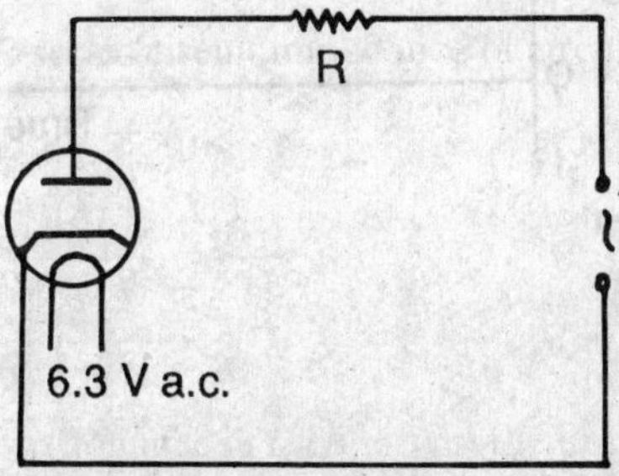

Which graph in figure given below represents the current through resistance ?

(a) A □
(b) B □

A

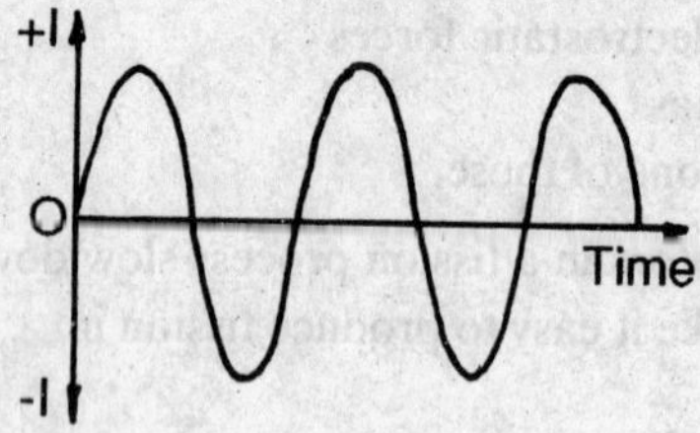

B

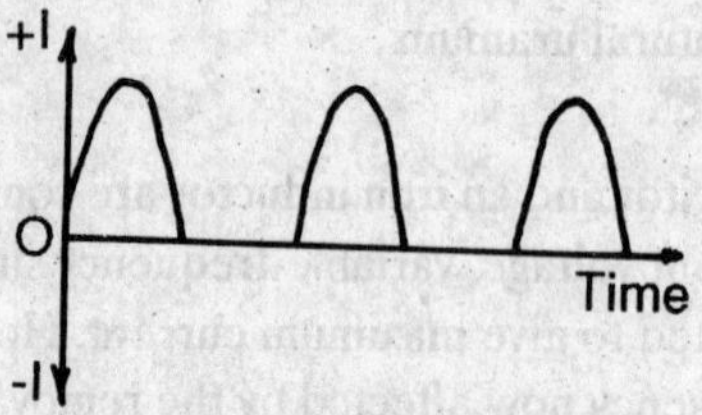

C

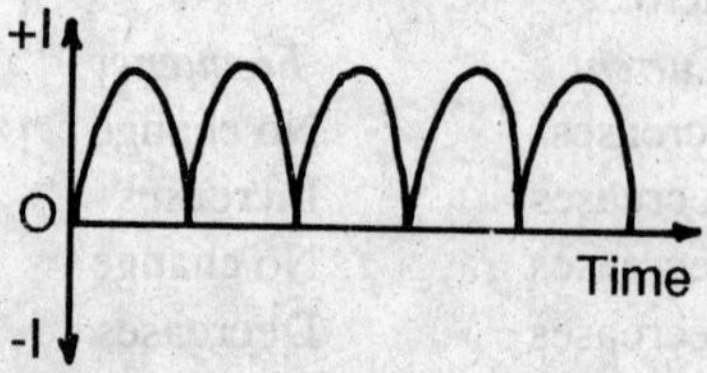

D

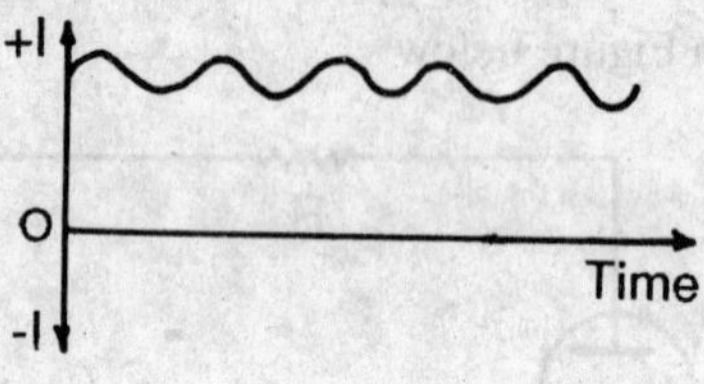

(*c*) C ☐
(*d*) D. ☐

Q. 228. In the circuit in Figure given below all four bulbs are identical. How do the brightnesses of the bulbs compare ?

(*a*) Bulb 1 is brightest ☐
(*b*) Bulb 4 is brightest ☐
(*c*) Bulbs 1 and 4 are brighter than 2 and 3 ☐
(*d*) Bulbs 3 and 4 are brighter than 1 and 2. ☐

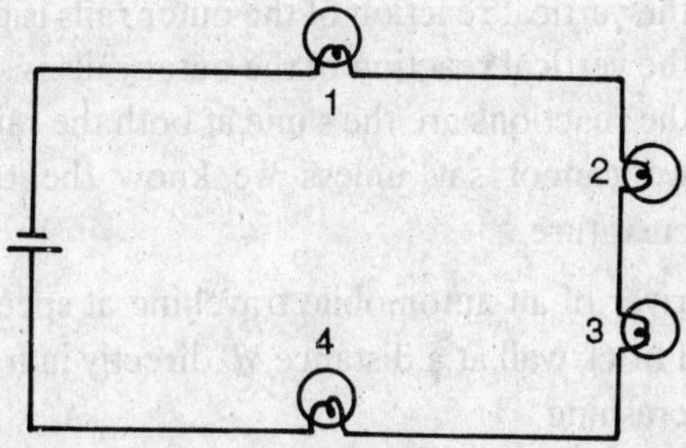

Q. 229. When a current of 2 A flows in a resistor for 10 minutes, 15,000 J of electrical energy is changed into heat energy. What is the voltage across the resistor ?

(*a*) $\frac{15,000}{2 \times 10 \times 60}$ ☐

(*b*) $\frac{15,000 \times 2}{10 \times 60}$ ☐

(*c*) $\frac{15,000}{2 \times 10}$ ☐

(*d*) $\frac{15,000}{10 \times 60}$ ☐

Q. 230. The temperature coefficient of resistance of a wire is 0.00125 per C, at 300 K its resistance is 1 ohm. The resistance of the wire will be 2 ohm at

(*a*) 1154 K ☐
(*b*) 1100 K ☐
(*c*) 1400 K ☐
(*d*) 1127 K. ☐

Q. 231. What is the relation between the voltage *V* across the ends of a resistor *R* and the current *I* which flows in *R*?

(*a*) $I = \frac{V}{R}$ ☐

(*b*) $V = \frac{I}{R}$ ☐

(*c*) $R = VI$ ☐

(*d*) $V = I^2R$. ☐

Q. 232. When a train moves fast on a horizontal curved track

(*a*) the vertical reaction of the outer rails is greater ☐
(*b*) the vertical reaction of the outer rails is smaller ☐
(*c*) the reactions are the same at both the rails ☐
(*d*) we cannot say unless we know the radius of curvature. ☐

Q. 233. The driver of an automobile travelling at speed '*v*' suddenly sees of brick wall at a distance '*d*' directly in front of him. To avoid crashing

(*a*) it is better to turn the car sharply away from the wall ☐
(*b*) it is better to slam on the brakes ☐
(c) the choice depends on the height of the brick wall ☐
(*d*) it is difficult to decide. ☐

Q. 234. A hunter aims his rifle with the barrel pointing at a monkey hanging from the branch of a tree. Just as the man fires, the monkey drops from the branch

(*a*) the bullet will pass above the monkey ☐
(*b*) the bullet will just miss the monkey ☐
(*c*) the monkey will be hit by the bullet ☐
(*d*) it depends on the acceleration due to gravity at the place. ☐

Q. 235. An object is attached to a vertical spring and it is slowly lowered to its equilibrium position. This stretches the spring by an amount '*d*'. If the same object is attached to the same vertical spring but permitted to fall instead, the string will be stretched through a distance

(*a*) $d/2$ ☐
(*b*) $3d/2$ ☐
(*c*) $1/2\, d^2$ ☐
(*d*) $2d$. ☐

Q. 236. The centre of mass of a system cannot change its state of motion, unless there is an external force acting on it. Yet the internal force of the brakes can bring a car to rest

(*a*) The brakes stop the wheels ☐
(*b*) The friction between the brake pads and the wheels stops the car ☐
(*c*) The car is stopped by the road ☐

(*d*) The car is stopped by the driver pressing the pedal. ☐

Q. 237. A ball is thrown vertically up and there is considerable air resistance. If t_1 is the time taken to rise and t_2 the time taken to fall

(*a*) $t_1 = t_2$ ☐

(*b*) $t_1 > t_2$ ☐

(*c*) $t_1 < t_2$ ☐

(*d*) The relationship depends on the ratio of air resistance to weight. ☐

Q. 238. A boy sitting in a rail-road car moving at constant velocity throws a ball straight up into the air. The car accelerates forward, while the ball is in the air

(*a*) The ball will fall behind him ☐

(*b*) The ball will fall in front of him ☐

(*c*) The ball will fall into his hand ☐

(*d*) The ball will fall slightly to the right. ☐

Q. 239. A body can have zero velocity and still be accelerating

(*a*) True ☐

(*b*) False ☐

(*c*) We cannot say unless the mass of the body is known ☐

(*d*) None of these. ☐

Q. 240. A man on the observation platform of a train moving with constant velocity drops a coin while leaning over the rail for *the man on the train*

(*a*) the path of the coin is a parabola ☐

(*b*) the path is a vertical straight line ☐

(*c*) the path is a straight line inclined to the horizontal ☐

(*d*) the path cannot be known unless the velocity of the train is known. ☐

Q. 241. Two spheres of the same volume but different densities are allowed to fall from the same height. They experience the same air resistance

(*a*) Both spheres reach the ground simulaneously ☐

(*b*) The heavier sphere reaches earlier ☐

(*c*) The heavier sphere reaches later ☐
(*d*) We cannot say unless the density of air is given. ☐

Q. 242. The direction of the velocity of a body cannot change, when its acceleration is constant. The statement is
(*a*) True ☐
(*b*) False ☐
(*c*) Uncertain ☐
(*d*) None of these. ☐

Q. 243. An elevator is descending at constant speed '*u*'. A passenger takes a coin from his pocket and drops it to the floor. For an observer at rest with respect to the elevator shaft, the acceleration of the coin is
(*a*) $u+g$ ☐
(*b*) $u-g$ ☐
(*c*) g ☐
(*d*) $2g$. ☐

Q. 244. A spring balance is laid horizontally on a table. A piece of string that is tied to one end of the spring balance passes over a smooth pulley at the edge of the table, and carries a 2 kg wt. A similar string tied to the other end of the balance passes over another pulley at the opposite edge of the table, and carries a 2 kg weight. The spring balance will read
(*a*) zero ☐
(*b*) 2 kg ☐
(*c*) 4 kg ☐
(*d*) some other value. ☐

Q. 245. The advantage of electron tube over samiconductor is
(*a*) its efficiency ☐
(*b*) its unlimited type ☐
(*c*) its law consumption lower ☐
(*d*) it takes no warming up time. ☐

Q. 246. When a triode is used as an amplifier, the input and output signals are
(*a*) out of phase by 180° ☐
(*b*) out of phase by 90° ☐
(*c*) out of phase by 45° ☐
(*d*) always in phase. ☐

Q. 247. Why is indirect beating of cathode surface preferred in vacuum tubes ?

(*a*) Because electron curission is uniform ☐
(*b*) Because design is simple ☐
(*c*) Because it is cheaper ☐
(*d*) Because it is the only way of obtaining electrons. ☐

Q. 248. In a common-emitter transistor circuit the mutual forward current ratis (*s*) is

(*a*) $\frac{\Delta I_e}{\Delta I_c}$ ☐
(*b*) $\frac{\Delta I_e}{\Delta I_c}$ ☐
(*c*) $\frac{\Delta I_e}{\Delta I_c}$ ☐
(*d*) $\frac{\Delta I_e}{\Delta I_c}$. ☐

Q. 249. In case of *p-n* junction diode at high valve of reverse bias the current rises sharply. This value of reverse bias is called

(*a*) Cut off voltage ☐
(*b*) Inverse voltage ☐
(*c*) Zener voltage ☐
(*d*) Critical voltage. ☐

Q. 250. In case of a triode valve the amplification it due to

(*a*) Anode ☐
(*b*) Cathode ☐
(*c*) Grid ☐
(*d*) All of these. ☐

Q. 251. What for is a semi-conductor diode used ?

(*a*) To convert d.c. to a.c. ☐
(*b*) To convert a.c. to d.c. ☐
(*c*) To increase voltage ☐
(*d*) To decrease voltage. ☐

Q. 252. On which of the following the cut off grid potential of a triode valve depends?

(*a*) Shafe of the grid ☐
(*b*) Grid potential ☐

(c) Plate potential
(d) All of these.

Q. 253. The correct relationship between μ, G_m and R_p, is

(a) $G_m \times R_p \times \mu = 1$
(b) $G_m = R_p \times \mu$
(c) $\mu = G_m \times R_p$
(d) $\mu = \frac{R_p}{G_m}$.

Q. 254. Which of the following relations given the value of junction resistance in case of a junction diode ?

(a) $\Delta E_f + \Delta I_g$
(b) $\Delta E_f / \Delta I_f$
(c) $\Delta I_f / \Delta E_f$
(d) $\Delta E_f - \Delta I_f$.

Q. 255. How is current plate effected by the introduction of a grid in a triode valve ?

(a) It increases plate potential
(b) It increases the electrons from the plate
(c) It helps in emission of electron at low temperature
(d) It neutralises space charge.

Q. 256. When a triode is used as an amplifier, it is operated in the

(a) curved portion of the characteristic curve
(b) straight portion of the characteristic curve
(c) bottom portion of the curve
(d) top of the curve.

Q. 257. The pressure inside electronic vacuum tube is of the order of

(a) zero
(b) 10^{-3} cm of Hg
(c) 10^{-7} cm of Hg
(d) 10^{-10} cm of Hg.

Q. 258. Why is a transistor preferred to a triode valve for use as amplifier ? This is because it

(a) can handle large power
(b) has a higher input impedence
(c) both the above
(d) None of these.

Q. 259. Diffraction is the property according to which light wave

(*a*) change their direction when they enter a different medium ☐
(*b*) bend round corners ☐
(*c*) produce same chemical effects ☐
(*d*) None of these. ☐

Q. 260. Raman effect is due to

(*a*) coherent scattering ☐
(*b*) incoherent scattering ☐
(*c*) no scattering ☐
(*d*) refraction. ☐

Q. 261. In order to measure the distance between two coherent sources in the bi-prism experiment, the convex lens should be placed

(*a*) between the source and the slit ☐
(*b*) between the slit and the bi-prism ☐
(*c*) between the bi-prism and the eye piece ☐
(*d*) None of these. ☐

Q. 262. Fresnel diffraction is a

(*a*) near field phenomenon ☐
(*b*) far field phenomenon ☐
(*c*) both of these ☐
(*d*) None of these. ☐

Q. 263. Grating law states that

(*a*) each slit in grating sends out a diffracted beam only ☐
(*b*) each slit in grating sends out a diffracted beam that interferes later to produce the pattern ☐
(*c*) each slit in the grating sends out an interfered beam only ☐
(*d*) $(a+b) = \frac{\pm n^{\lambda}}{\sin\theta}$ for the first order principal maxima. ☐

Q. 264. Rowland's concave grating produces a spectrum which is

(*a*) free from spherical aberration only ☐
(*b*) free form chromatic aberration only ☐
(*c*) free from chromatic and other aberrations ☐

(*d*) free from distortions. ☐

Q. 265. Through a soap film different colours are seen by white light due to which one of the following is

(*a*) dispersion ☐
(*b*) diffraction ☐
(*c*) interference ☐
(*d*) reflection. ☐

Q. 266. Fermat's principle states that

(*a*) light chooses random path ☐
(*b*) light bends round corners ☐
(*c*) light chooses a path of least time ☐
(*d*) None of these. ☐

Q. 267. Longitudinal chromatic aberration is

(*a*) dispersive power/mean focal length ☐
(*b*) mean focal length ☐
(*c*) dispersive power ☐
(*d*) dispersive power × mean focal length. ☐

Q. 268. According to Huygen's principle

(*a*) light bends round corners ☐
(*b*) light travels in a straight line ☐
(*c*) all points on primary wave front are considered to be centre of disturbances ☐
(*d*) None of these. ☐

Q. 269. Which of the following can be used to remove spherical aberration in a spherical mirror ?

(*a*) Limiting the number of rays by using a stop ☐
(*b*) Using an ellipsoidal mirror ☐
(*c*) Using a parabolic mirror ☐
(*d*) Using any of the above. ☐

Q. 270. The condition that must be satisfied in order that a combination of two lenses of focal length f_1 and f_2 at a distance '*d*' cm apart be chromatic is

(*a*) $d = (f_1 + f_2)$ ☐
(*b*) $d = \frac{f_1 + f_2}{2}$ ☐
(*c*) $d = (f_1 - f_2)$ ☐

(*d*) None of these. ☐

Q. 271. A double convex air bubble in water will behave as

(*a*) plane slab ☐
(*b*) concave mirror ☐
(*c*) convex lens ☐
(*d*) concave lens. ☐

Q. 272. A particle having rest mass equal to that of electron but charge equal and opposite to that of electron is called

(*a*) proton ☐
(*b*) positron ☐
(*c*) hyperon ☐
(*d*) meson. ☐

Q. 273. Which is the most massive particle ?

(*a*) Deutron ☐
(*b*) α-particle ☐
(*c*) Neutron ☐
(*d*) Positron. ☐

Q. 274. Which of the following is believed to be carrier of nuclear forces ?

(*a*) Lepton ☐
(*b*) Meson ☐
(*c*) Bason ☐
(*d*) Baryon. ☐

Q. 275. Which of the following particles never decays into space ?

(*a*) μ-meson ☐
(*b*) Hyperon ☐
(*c*) Proton ☐
(*d*) Neutron. ☐

Q. 276. Protons are accelerated in one step by which of the following acceleration ?

(*a*) Van Graff acceleration ☐
(*b*) Cyclotron ☐
(*c*) Spark chamber ☐
(*d*) Linear accelerator. ☐

Q. 277. Electron volt (*eV*) is the unit of

(*a*) potential ☐

(*b*) electric field ☐
(*c*) energy ☐
(*d*) charge. ☐

Q. 278. Who was the first to suggest the existence of a positively charged nucleus in an atom ?
(*a*) J.J. Thomson ☐
(*b*) R.A. Millikan ☐
(*c*) E.Rutherford ☐
(*d*) N. Bohr. ☐

Q. 279. Which of the following can be used as an arrester in a nuclear reactor ?
(*a*) Graphite ☐
(*b*) Heavy water ☐
(*c*) Uranium ☐
(*d*) Cadmium. ☐

Q. 280. The function of an arrester in a nuclear reactor is to
(*a*) aid in the fast multiplication of star moving electrons ☐
(*b*) slow down fast neutrons to avoid non-radiative capture ☐
(*c*) absorbs the slow moving neutrons ☐
(*d*) decreases the fission cross-section of the neutrons. ☐

Q. 281. Which of the following may be considered as clean bomb (those which do not spread radioactivity) ?
(*a*) Atom bomb ☐
(*b*) Hydrogen bomb ☐
(*c*) Neutron bomb ☐
(*d*) Both (*b*) and (*c*). ☐

Q. 282. Which is an example of nuclear fusion ?
(*a*) Formation of water from hydrogen and oxygen ☐
(*b*) Formation of barium and krypton from uranium ☐
(*c*) Formation of helium from hydrogen ☐
(*d*) None of these. ☐

Q. 283. Who was awarded noble prize for predicting meson ?
(*a*) Yukawa ☐

(*b*) Pauli

(*c*) Schordinger

(*d*) None of these.

Q. 284. Who was awarded Nobel Prize for first artificial atomic transmutation ?

(*a*) Yukawa

(*b*) Rutherford

(*c*) Chadwick

(*d*) None of these.

Q. 285. In Carnot cycle

(*a*) work done during adiabatic expansion is less than work done during adiabatic compression

(*b*) work done by working substance during adiabatic expansion is greater than work done during adiabatic compression

(*c*) work done during adiabatic expansion is equal to work done during adiabatic compression

(*d*) work done during adiabatic expansion is equal to the heat absorbed from the source.

Q. 286. If T_1 is the temperature of the source, and T_2 temperature of the sink, in a Carnot engine the efficiency is

(*a*) $\frac{T_1 - T_2}{T_2}$

(*b*) $\frac{T_1 + T_2}{T_2}$

(*c*) $\frac{T_1 - T_2}{T_1}$

(*d*) $\frac{T_1 + T_2}{T_1}$.

Q. 287. If ρ is the adiabatic compression ratio, and γ the ratio of specific heats, the efficiency of the carnot engine is

(*a*) $1-\rho^{\gamma}$

(*b*) $1-\left(\frac{1}{\rho}\right)^{\gamma-1}$

(*c*) $1+\rho^{\gamma}$

(*d*) $1-\rho^{-\gamma}$.

Q. 288. Kelvin's statement of the second law of thermodynamics is called law of Degradation of Energy because

(*a*) all the heat supplied to the working substance is converted into work ☐

(*b*) in heat engines, heat can be converted to mechanical energy ☐

(*c*) some of the heat supplied to the working of substance is not available ☐

(*d*) the heat engine is not very efficient. ☐

Q. 289. Carnot cycle is known as constant temperature cycle because

(*a*) heat is taken in at constant lower temperature ☐

(*b*) temperature remains constant throughout the cycle ☐

(*c*) heat is rejected at constant higher temperature ☐

(*d*) heat is taken in at constant temperature and rejected at constant temperature. ☐

Q. 290. In practice a diesel engine can have a greater operating efficiency because

(*a*) the efficiency of the Otto engine decreases after a certain compression ratio ☐

(*b*) for the same compression ratio $\eta_d < \eta_0$ ☐

(*c*) the compression ratio of a diesel engine value than for an Otto engine ☐

(*d*) the presence of the spark plug lowers efficiency. ☐

Q. 291. Thermodynamics is a subject dealing with

(*a*) the motion of molecules in matter ☐

(*b*) the motion of atoms in matter ☐

(*c*) the macroscopic variables, such as pressure, temperature and volume ☐

(*d*) the kinetic energy of molecules. ☐

Q. 292. The temperature of a gas is related to

(*a*) the total kinetic energy of the molecules of the gas ☐

(*b*) the kinetic energy of the centre of mass ☐

(*c*) the potential of energy of the centre of mass ☐

(*d*) the total kinetic energy of the molecules with respect to centre of mass of the gas. ☐

Q. 293. The average kinetic energy of a molecule in a gas at temperature $T^o K$ is

(*a*) $\frac{1}{3}kT$ ☐

(*b*) $\frac{3}{2}kT$ ☐

(*c*) $\frac{1}{2}kT$ ☐

(*d*) $\frac{2}{3}kT$. ☐

Q. 294. Mean free path in a gas is

(*a*) the distance travelled by a molecule before hitting a wall ☐

(*b*) the average distance travelled by a molecule in one second ☐

(*c*) the average distance travelled by molecules between any two successive collisions ☐

(*d*) the root mean square velocity. ☐

Q. 295. Viscosity in a gas is explained by

(*a*) force of attraction between the molecules ☐

(*b*) collision between molecules ☐

(*c*) transport of kinetic energy from layer ☐

(*d*) transport of momentum from one layer to another. ☐

Q. 296. For an atom of hydrogen the radius of first orbit (γ_0) is given by

(*a*) $\gamma_0 = \frac{h}{me^2}$ ☐

(*b*) $\gamma_0 = \frac{me}{4h^2}$ ☐

(*c*) $\gamma_0 = \frac{h^2}{4\pi^2 me^2}$ ☐

(*d*) $\gamma_0 = h^2 . m . e^2$. ☐

Q. 297. Which of the following particles leave no trail in a cloud chamber ?

(*a*) Electron ☐

(b) Proton
(c) α-particle
(d) Neutron.

Q. 298. Which of the following is formed by decay of a free neutron ?
(a) A number of electrons
(b) Two protons
(c) A proton and an electron
(d) An α-particle.

Q. 299. Select the pair that represents Isotopes
(a) ${}_{37}X^{75}$ and ${}_{37}X^{75}$
(b) ${}_{37}X^{75}$ and ${}_{37}X^{77}$
(c) ${}_{37}X^{75}$ and ${}_{35}X^{75}$
(d) ${}_{37}X^{75}$ and ${}_{35}X^{75}$.

Q. 300. In an atom of hydrogen if the radius of first orbit is 0.528 Å then the radius of 4th orbit is approximately
(a) 8.45 Å
(b) 4.85 Å
(c) 5.84 Å
(d) 0.845 Å.

Section IV

Q. 301. The maximum kinetic energy of photoelectrons emitted by a photosensitive metallic surface depends on
(a) intensity of incident radiation
(b) frequency of incident radiation
(c) Both the above
(d) None of these.

Q. 302. Which of the following is used as a fuel in a nuclear reactor?
(a) Water/heavy water
(b) Lead/aluminium
(c) Lead/cadmium
(d) Plutonium/uranium.

Q. 303. Photoelectric cell is a device that converts
(*a*) electrical energy into light energy ☐
(*b*) electrical energy into sound energy ☐
(*c*) light energy into electrical energy ☐
(*d*) None of these ☐

Q. 304. Which of the following constitute β-particles ?
(*a*) Electrons ☐
(*b*) Protons ☐
(*c*) Photons ☐
(*d*) None of these. ☐

Q. 305. A positron is
(*a*) a helium nucleus ☐
(*b*) made up of a neutron and two protons ☐
(*c*) an antiparticle of electron ☐
(*d*) None of these. ☐

Q. 306. Which of the following terms is used for a process in which the nucleus of a process in which the nucleus of a heavy atom breaks into two more or less equal segments accompanied by a release of large amount of energy ?
(*a*) Fusion ☐
(*b*) Fission ☐
(*c*) Ionisation ☐
(*d*) None of these. ☐

Q. 307. Why does nuclear fusion requires high temperature ?
(*a*) Because all nuclear reactions absorbs heat ☐
(*b*) The mass deficit must be supplied ☐
(*c*) The particles can not come together unless they move rapidly ☐
(*d*) The binding energy must be supplied from some external sources. ☐

Q. 308. Who discovered radioactivity ?
(*a*) M. Curie ☐
(*b*) Becqueral ☐
(*c*) Both the above ☐
(*d*) None of these. ☐

Q. 309. Which of the following tends to make a nucleus fly apart?

(*a*) Nuclear force ☐
(*b*) Magnetic force ☐
(*c*) Gravitational force ☐
(*d*) Electrostatic repulsion ☐

Q. 310. The source of energy of the sun is
(*a*) fusion of heavy nuclei ☐
(*b*) fusion of light nuclei ☐
(*c*) fission of light nuclei ☐
(*d*) All the above. ☐

Q. 311. Rejector circuits are
(*a*) parallel resonant circuits ☐
(*b*) series resonant circuits ☐
(*c*) Both the above ☐
(*d*) None of these. ☐

Q. 312. If a substance is placed in a watch glass on the pole pieces of a magnet, the substance accumulates on the sides causing a depression at the centre. It may be concluded that the substance is
(*a*) ferromagnetic in nature ☐
(*b*) paramagnetic in nature ☐
(*c*) diamagnetic in nature ☐
(*d*) None of these. ☐

Q. 313. The spacing of electric field lines between two idential point charges of opposite sign
(*a*) is not dependent on the magnitude on the charges ☐
(*b*) is an indication of the field direction ☐
(*c*) is an indication of the field strength ☐
(*d*) is large when the charges are very large. ☐

Q. 314. As a positively-charged rod is brought closer and closer to a positively charged electroscope, the gold leaf
(*a*) diverges ☐
(*b*) converges ☐
(*c*) is neutralized ☐
(*d*) is unaffected. ☐

Q. 315. According to Gauss law the number of electric field lines crossing any closed surface is

(*a*) numerically equal to the enclosed charge ☐
(*b*) equal to the enclosed positive charge ☐
(*c*) equal to the electric field inside the surface ☐
(*d*) equal to the charge density on the surface. ☐

Q. 316. The permitivity of a medium
(*a*) is a measure of its density ☐
(*b*) is equal to unity for air or a vacuum ☐
(*c*) is dependent on charge density of the medium ☐
(*d*) determines the magnitude of an electric field that can be establish by the medium. ☐

Q. 317. In the A.C. circuit below the supply has fixed voltage an a variable frequency e.m.f of the source is

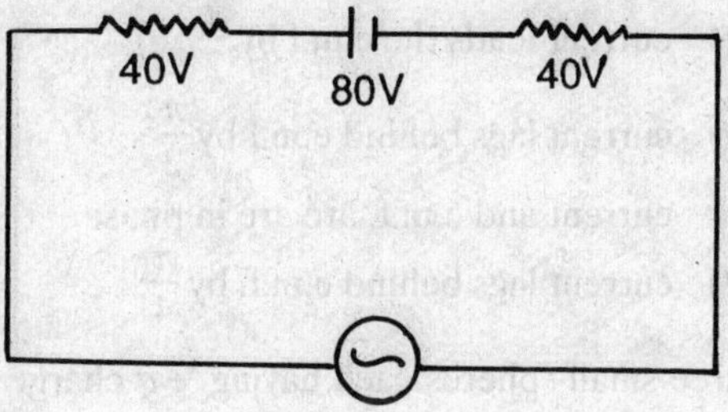

(*a*) 0 ☐
(*b*) 100 ☐
(*c*) 250 ☐
(*d*) $40\sqrt{2}$. ☐

Q. 318. A charged particle moving at right angles to a uniform magnetic field follows
(*a*) circular path ☐
(*b*) parabolic path ☐
(*c*) straight path ☐
(*d*) elliptical path. ☐

Q. 319. The ratio of intensity of magnetisation to the magnetic force is known as
(*a*) permeability ☐
(*b*) magnetic induction ☐
(*c*) magnetic intensity ☐
(*d*) magnetic susceptibility. ☐

Q. 320. Electromagnetic induction is involved in

(*a*) the charging of a storage battery ☐

(*b*) the operation of an electromagnet ☐

(*c*) the production of current by relative motion between a magnet and a wire ☐

(*d*) the charging of an electroscope without actually touching it with a charged rod. ☐

Q. 312. Cyclotron is a

(*a*) liner accelerator ☐

(*b*) circular accelerator ☐

(*c*) rectangular accelerator ☐

(*d*) None of these. ☐

Q. 322. In a circuit containing capacitance only

(*a*) current leads the e.m.f by $\frac{\pi}{2}$ ☐

(*b*) current lags behind e.m.f by $\frac{\pi}{2}$ ☐

(*c*) current and e.m.f. are are in phase ☐

(*d*) current lags behind e.m.f. by $\frac{\pi}{4}$. ☐

Q. 323. Three small spheres each having $+q$ charge are kept on the circumference of a circle in such a way that they make an equilateral triangle. The electric intensity at the centre of the circle is

(*a*) zero ☐

(*b*) $\frac{1}{4\pi\varepsilon_0} \cdot \frac{q}{r^2}$ ☐

(*c*) $\frac{1}{4\pi\varepsilon_0} \cdot \frac{2q}{r^2}$ ☐

(*d*) $\frac{1}{4\pi\varepsilon_0} \cdot \frac{q}{r^2}\sqrt{3}$. ☐

Q. 324. Which of the diagrams in Figure below illustrates the effect of bringing a negatively charged rod near an earthed electroscope ?

(*a*) *A* ☐

(*b*) *B* ☐

(*c*) *C* ☐

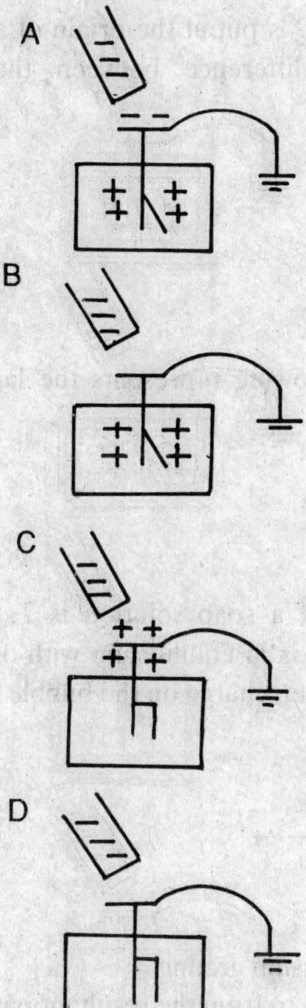

(*d*) D. ☐

Q.325. The E-field at a distance '*r*' from the infinite line of charge of 'λ' linear charge density is given by

(*a*) $E = \dfrac{\lambda}{4\pi\varepsilon_0 r}$ ☐

(*b*) $E = \dfrac{\lambda}{2\pi\varepsilon_0 r^2}$ ☐

(*c*) $E = \dfrac{\lambda}{2\pi\varepsilon_0 r^2}$ ☐

(*d*) $E = \dfrac{\lambda}{2\pi\varepsilon_0 r}$. ☐

Q. 326. A charge of 20 μC is put at the origin of an x-y coordinate. Then potential difference between the points $(a, 0)$ $\left(\frac{a}{\sqrt{2}}, \frac{a}{\sqrt{2}}\right)$ is

(a) $\frac{a}{\sqrt{2}}$ ☐

(b) a ☐

(c) zero ☐

(d) $2a$. ☐

Q. 327. Which of the following represents the largest measure of charge

(a) 100 μC ☐

(b) 10 μC ☐

(c) 10^{12} electron ☐

(d) 10^{-7} C. ☐

Q. 328. Surface tension of a soap solution is T. A charged soap bubble of radius r is in equilibrium with outside and inside pressure equal. Then charge on the bubble is

(a) $4\pi\gamma^2 \frac{\sqrt{8T\varepsilon_0}}{\gamma}$ ☐

(b) $4\pi\gamma \sqrt{8T.\varepsilon_0}$ ☐

(c) $4\pi\gamma^2 \sqrt{8T \,.\, \varepsilon_0 \,.\, \gamma^2}$ ☐

(d) $4\pi\gamma^2 \quad \frac{\gamma}{8T \,.\, \varepsilon_0}$. ☐

Q. 329. In a plane transmission grating

(a) light diffracts to from the resultant patterns ☐

(b) light interferes to from the resultant patterns ☐

(c) light polarises to from the resultant patterns ☐

(d) light diffracts and interferes to from resultant patterns. ☐

Q. 330. If one looks at the filament of an incandescent lamp at a distance, through a fine silk handkerchief, a central white filament with several coloured filaments on the side is seen. It is an example of

(a) thin film interference ☐

(b) thick film interference ☐

(c) rotatory polarisation ☐

(d) diffraction. ☐

Q. 331. Dispersive power of a granting is given by

(*a*) $\frac{d\theta}{d\lambda} = \frac{m\lambda}{\sin\theta}$ ☐

(*b*) $\frac{d\theta}{d\lambda} = \frac{m\ .\ x}{\cos\theta}$ ☐

(*c*) $\frac{d\theta}{d\lambda} = \frac{m\ .\ x}{\tan\theta}$ ☐

(*d*) $\frac{d\theta}{d\lambda} = \frac{m\lambda}{\lambda\sin\theta}$ ☐

Q. 332. The number of lines needed on a grating to just resolve 4000 and 4001 Å in the first order spectrum is

(*a*) 4000 ☐
(*b*) 8000 ☐
(*c*) 2000 ☐
(*d*) 400 ☐

Q. 333. Stars are not visible in day light because

(*a*) the stars vanish during the day ☐
(*b*) the stars hide behind the sun ☐
(*c*) Both (*a*) and (*b*) ☐
(*d*) the air molecules of the atmosphere scatter the sun light into a blanket of strong light through which the faint stars cannot be seen. ☐

Q. 334. LASER is a device for

(*a*) producing a beam of white light ☐
(*b*) producing a beam of monochromatic an coherent light ☐
(*c*) producing a beam of high intensity incoherent light ☐
(*d*) producing highly penetrating X-rays. ☐

Q. 335. A laser beam may be used to measure very large distance because, it

(*a*) is unidirectional ☐
(*b*) is coherent ☐
(*c*) is monochromatic ☐
(*d*) is not absorbed. ☐

Q. 336. Sky appears blue because

(*a*) there is more blue colour is sunlight than any other colour ☐

(*b*) short waves are scattered more than the long waves by the atmosphere ☐

(*c*) the eye is more sensitive to blue ☐

(*d*) the atmosphere absorbs long wavelenghts more than short wavelengths. ☐

Q. 337. The wave front is an imaginary surface at which

(*a*) constant phase difference is always maintained ☐

(*b*) phase is always same for all points ☐

(*c*) phase changes with constant rate in all directions along the surface ☐

(*d*) phase changes with the rate $2\pi/\lambda$ per unit length in all directions along the surface. ☐

Q. 338. When sunlight enters earth's atmosphere the rays which get scattered the most are

(*a*) violet and blue ☐

(*b*) yellow and green ☐

(*c*) red and orange ☐

(*d*) white and blue. ☐

Q. 339. The phenomenon of fluorescence differs from the Raman scattering in one of the following ways.

(*a*) Intensity of Raman lines is of the same order as that of fluorescent lines ☐

(*b*) The reradiated energy in the case of fluorescence is of a frequency other than the one which the fluorescing substance is able to absorb ☐

(*c*) The intensity of Raman lines is of an entirely different order than that of fluorescent lines ☐

(*d*) None of these. ☐

Q. 340. Which of the following phenomenon is responsible for the production of shadow ?

(*a*) Interference ☐

(*b*) Diffraction ☐

(*c*) Polarization ☐

(*d*) Rectilinear propagation of light. ☐

Q. 341. A light beam is said to be plane polarised when

(*a*) its vibrations are restricted to only one plane ☐
(*b*) its vibrations are very strong in one plane ☐
(*c*) its vibrations take place in any plane ☐
(*d*) None of these. ☐

Q. 342. Which of the following was confirmed by the discovery of polarization of light ?

(*a*) Light waves are transverse in nature ☐
(*b*) Light waves are longitudinal in character ☐
(*c*) Light waves are neither waves nor particles ☐
(*d*) All the above. ☐

Q. 343. The defect in image due to oblique centric rays falling on the lens is called

(*a*) astigmatism ☐
(*b*) coma ☐
(*c*) curvature of image field ☐
(*d*) spherical aberration. ☐

Q. 344. The blurring of the image due to dispersion in lens is called

(*a*) spherical aberration ☐
(*b*) chromatic aberration ☐
(*c*) astigmatism ☐
(*d*) curvature of image field. ☐

Q. 345. The difficulty generally experienced while using a Huygen's eye-piece is that

(*a*) intensity of image is poor ☐
(*b*) the image is too much faint w.r.t. scale ☐
(*c*) the image is much illuminated w.r.t. scale ☐
(*d*) magnification of image and cross-wires are different. ☐

Q. 346. In case of telescope, larger the diameter of the objective

(*a*) greater its resolving power ☐
(*b*) lesser its resolving power ☐
(*c*) lesser its magnifying power ☐
(*d*) greater its magnifying power. ☐

Q. 347. The function of field lens of an eyepiece is

(*a*) field lens with relatively large aperture helps increase the field of view of the observer ☐

(*b*) field lens in conjugation with eye-lens reduces spherical aberration ☐
(*c*) both (*a*) and (*b*) ☐
(*d*) eye-piece comprises of only one lens. ☐

Q. 348. A direct vision spectroscope is based upon
(*a*) deviation without dispersion ☐
(*b*) dispersion without deviation ☐
(*c*) no deviation and no dispersion ☐
(*d*) Any of the above. ☐

Q. 349. A direct vision spectroscope is better than a prism spectrometer because it helps in
(*a*) observing the spectrum without a source ☐
(*b*) observing the spectrum perpendicular to a source ☐
(*c*) observing the spectrum in line with the source ☐
(*d*) None of these. ☐

Q. 350. When we decrease the diameter of the objective lens of a telescope, then the resolution of telescope
(*a*) decreases ☐
(*b*) increases ☐
(*c*) remains the same ☐
(*d*) depends on the focal length of the lens. ☐

Q. 351. The resolving power of an instrument increases as the wavelength of light used decreases, then the magnifying power will
(*a*) remain constant ☐
(*b*) decrease ☐
(*c*) increase ☐
(*d*) no relation between the two. ☐

Q. 352. The resolving power of a grating can be represented as
(*a*) $N \times n$ ☐
(*b*) N^2/n ☐
(*c*) n/N ☐
(*d*) $(N \times n)^2$. ☐

(where N is the total number of lines on the grating and n is the order of the spectrum for given wavelength)

Q. 353. The limit of resolution of microscope can be correctly given by

(*a*) $1.22.\lambda N.A$ ☐

(*b*) $\dfrac{\lambda^2}{1.22\, N.A.}$ ☐

(*c*) $\dfrac{122\lambda}{2N.A.}$ ☐

(*d*) $\dfrac{\lambda}{1.22\, N.A.}$ ☐

(where N.A. = numerical aperature)

Q. 354. For a prism of particular material and given wavelength λ, the resolving power varies as

(*a*) first power of length of its base ☐

(*b*) square of inverse of length of its base ☐

(*c*) increase of length of its base ☐

(*d*) cube of the length of its base. ☐

Q. 355. Epidiascope is made by combining

(*a*) a telescope and a microscope ☐

(*b*) a binocular and a telescope ☐

(*c*) a projection lantern and compound microscope ☐

(*d*) a projection lantern and a simple microscope. ☐

Q. 356. A man is standing on a spring scale in an elevator. When the elevator is at rest, scale reads 160 kg. When the elevator starts to move, the scale reads 140 kg. The elevator has

(*a*) an upward acceleration ☐

(*b*) a constant upward speed ☐

(*c*) a constant downward speed ☐

(*d*) downward acceleration. ☐

Q. 357. A 1 kg mass has a kinetic energy of 1 joule when its speed is

(*a*) 1.4 m/sec ☐

(*b*) 0.45 m/sec ☐

(*c*) 1 m/sec ☐

(*d*) 4.4 m/sec. ☐

Q. 358. The algebraic sign of acceleration depends on

(*a*) the choice of direction ☐

(*b*) the sign of the final velocity ☐

(c) whether an object is speeding up or slowing down
(d) the position of the object.

Q. 359. A solid iron cylinder A rolls down an inclined plane, while an identical sphere B slides down the plane in a frictionless manner. At the bottom the kinetic energy of sphere A is
(a) equal to that of sphere B
(b) more than that of sphere B
(c) less than that of sphere B
(d) more or less than that of sphere B, depending upon the angle of the plane.

Q. 360. When a body is accelerated
(a) its velocity always changes
(b) its direction always changes
(c) its speed always changes
(d) its falls towards the earth.

Q. 361. A 2000 Kg truck travelling at 36 km/hr strikes a tree and comes to a stop in 0.1 sec. The average force on the truck during the crash is
(a) 2×10^3 newtons
(b) 2×10^2 newtons
(c) 2×10^4 newtons
(d) 2×10^5 newtons.

Q. 362. A thief stole a box of weight W and while carrying on his head jumped down from third storey of the building. Before he reached the ground, he experienced a load of
(a) zero
(b) less than W
(c) infinite
(d) more than W.

Q. 363. If the initial velocity, the distance travelled, and the time elapsed are known, which equation would you use to calculate the acceleration ?
(a) $v_f = v_o + at$
(b) $s = v_o t + \frac{1}{2}at^2$
(c) $s = vt$
(d) $2as = v_f - v_o$.

Q. 364. Ampere's law is concerned with
(*a*) the force on a wire carrying a current in a magnetic field ☐
(*b*) electrochemical equivalent ☐
(*c*) unit magnetic poles ☐
(*d*) r.m.s. values. ☐

Q. 365. "The direction of an induced current is always such that the magnetic field on it tends to oppose the charge in the strength of magnetic filed belonging to primary current" is the statement for
(*a*) Faraday's law ☐
(*b*) Ampere's law ☐
(*c*) Henry's law ☐
(*d*) Lenz's law. ☐

Q. 366. When a charged particle moves through a magnetic field, it suffers a change in its
(*a*) charge ☐
(*b*) energy ☐
(*c*) mass ☐
(*d*) direction of motion. ☐

Q. 367. Lenz's law follows from the law of conservation of
(*a*) mass ☐
(*b*) charge ☐
(*c*) energy ☐
(*d*) momentum. ☐

Q. 368. According to Faraday's law, the total charge induced in a conductor that is moved in a magnetic field depends upon
(*a*) initial magnetic flux ☐
(*b*) final magnetic flux ☐
(*c*) rate of change of magnetic flux ☐
(*d*) change in magnetic flux. ☐

Q. 369. According to Faraday's law, the total charge induced in a conductor that is moved in magnetic field depends upon
(*a*) initial magnetic flux ☐
(*b*) final magnetic flux ☐
(*c*) rate of change of magnetic flux ☐
(*d*) change in magnetic flux. ☐

Q. 370. The direction of the induced e.m.f. during electromagnetic induction can be determined by making use of

(*a*) Faradays law ☐
(*b*) Amperes law ☐
(*c*) Lenz's law ☐
(*d*) Laplace law. ☐

Q. 371. Which of the following figures shows the variation of current with time in a tungsten filament lamp, from the moment current flows ?

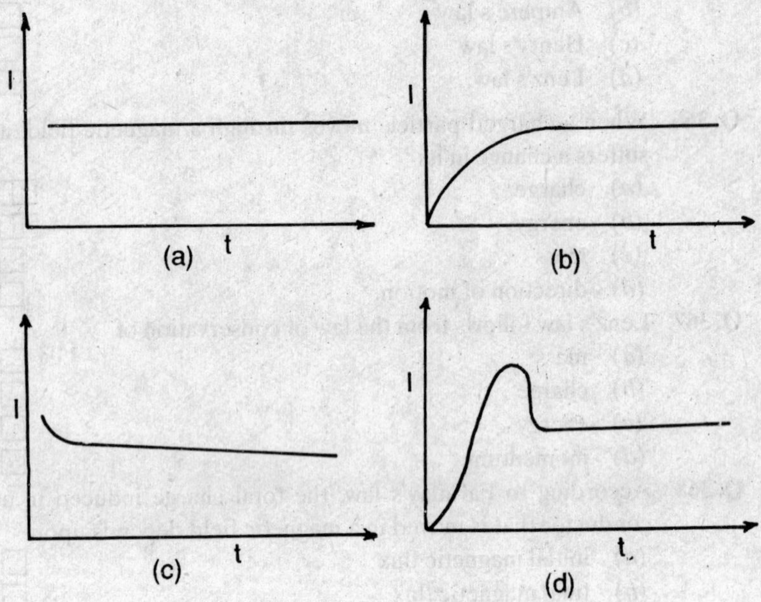

Q. 372. If E is the coefficient of volume elasticity of the medium, ρ its density, then the velocity of a longitudinal wave in a fluid is given by

(*a*) $v = E/\rho$ ☐
(*b*) $v = \sqrt{E/\rho}$ ☐

(c) $v = \rho / E$ ☐

(d) $v = \sqrt{\rho / E}$ ☐

Q. 373. Ultrasonic sound waves

(a) can be heard by a normal human ear ☐

(b) can be heard with the help of hearing apparatus ☐

(c) cannot be heard ☐

(d) can be heard with the help of microphone. ☐

Q. 374. If a car is moving towards you with its horn sounding, you hear a higher pitch than you would hear if the car were at rest, this is because

(a) the waves travel faster ☐

(b) successive crests produced by the horn are closer to each other ☐

(c) you received successive crests more frequently even though the wave itself is unchanged ☐

(d) the motion of the car raises the frequency of the horn. ☐

Q. 375. The quality of a tone

(a) decreases with loudness ☐

(b) varies directly as pitch ☐

(c) varies inversely as its frequency ☐

(d) depends upon the overtones that are present there. ☐

Q. 376. Two particles are executing S.H.M. of same period, T. If the second particle starts from its mean position $T/2$ later than the first, the phase difference between the two particles at any instant of time is

(a) $\pi/2$ ☐

(b) π ☐

(c) $2\pi/2$ ☐

(d) 2π. ☐

Q. 377. A particle moving with S.H.M. passes through two points A and B, with the same velocity, having occupied 2 seconds in passing from A to B ; after another 2 seconds it returns to B. The period is

(a) 2 seconds ☐

(b) 4 seconds ☐
(c) 5 seconds ☐
(d) 8 seconds. ☐

Q. 378. A tuning fork A produces 4 beats with a tuning fork B of frequency 256. A is filled at the top of the prong and the number of beats increases. What was its original frequency ?

(a) 260 ☐
(b) 258 ☐
(c) 252 ☐
(d) 250. ☐

Q. 379. Point out the wrong statement. In a longitudinal stationary wave

(a) every particle excepting nodes executes SHM ☐
(b) the amplitude varies continuously along the line assuming +ve, zero and −ve values ☐
(c) nodes are the points where the strain is zero ☐
(d) antinodes are the points where the displacement is maximum. ☐

Q. 380. A rod clamped at the midpoint can be excited to produce longitudinal stationary waves. The frequency is given by $n = \frac{s}{2L} \quad \frac{Y}{P}$ In this expression

(a) s is the number of loops ☐
(b) s is the number of antinodes ☐
(c) s is the number of modes ☐
(d) s is the number of nodes and anitnodes. ☐

Q. 381. A tuning fork is a case of

(a) a rod with longitudinal vibrations ☐
(b) two fixed free rods with transverse vibrations ☐
(c) vibrations of a rod rich in overtones ☐
(d) frequency being independent of temperature. ☐

Q. 382. The permittivity of a medium

(a) is a measure of its density ☐
(b) is equal to unity for air/vacuum ☐
(c) is dependent on charge density of medium ☐
(d) determines the magnitude of an electric field that can be established by the medium. ☐

Q. 383. When a wire moves perpendicular to a magnetic field, the induced e.m.f. does not depend upon

(*a*) the velocity of the wire ☐
(*b*) the resistance of the wire ☐
(*c*) the flux density of the magnetic field ☐
(*d*) the orientation of the wire. ☐

Q. 384. Neutral temperature of a thermocouple is the temperature at which

(*a*) the thermo e.m.f. changes sign ☐
(*b*) the thermo e.m.f. is maximum ☐
(*c*) the thermo e.m.f. is minimum ☐
(*d*) the thermo current is maximum. ☐

Q. 385. Under which of the following conditions current in a circuit is wattless ?

(*a*) When inductance in circuit is zero ☐
(*b*) When resistance in circuit is zero ☐
(*c*) when current is alternating ☐
(*d*) When resistance and inductance both are zero. ☐

Q. 386. In which of the following, the loss of energy is less ?

(*a*) D.C. ☐
(*b*) A.C. ☐
(*c*) Thermoelectricity ☐
(*d*) Photoelectricity. ☐

Q. 387. In a *LCR* series circuit impedance of circuit is

(*a*) $R + X_L + X_C$ ☐
(*b*) $R + (X_L - X_C)$ ☐
(*c*) $\sqrt{R^2 + (X_L^2 - X_C^2)}$ ☐
(*d*) $\sqrt{R^2 + (X_L - X_C)^2}$. ☐

Q. 388. At series resonance in *LCR* circuit the impedance is equal to

(*a*) ohmic resistance ☐
(*b*) inductive reactance ☐
(*c*) capacitive reactance ☐
(*d*) inductive reactance minus capacitive reactance. ☐

Q. 389. According to Gauss' law, the number of electric field lines crossing the closed surface is

(a) numerically equal to the enclosed charge ☐
(b) equal to the enclosed positive charge ☐
(c) equal to the electric field inside the surface ☐
(d) equal to the charge density on the surface. ☐

Q. 390. In what form is the energy stored in an inductor ?
(a) Magnetic ☐
(b) Electrostatic ☐
(c) Both magnetic and electrostatic ☐
(d) None of these. ☐

Q. 391. The product of absolute permittivity and absolute permeability of free space is given by
(a) C^2 ☐
(b) C ☐
(c) $\frac{1}{C}$ ☐
(d) $\frac{1}{C^2}$. ☐

Q. 392. In what form is the energy stored in an capacitor ?
(a) Magnetic ☐
(b) Electrostatic ☐
(c) Both magnetic and electrostatic ☐
(d) None of these. ☐

Q. 393. If a paramagnetic liquid is placed in a watch glass resting on the pole pieces, the liquid accumulates
(a) where the field is strongest ☐
(b) where the field is weak ☐
(c) is not affected ☐
(d) where the field is zero. ☐

Q. 394. Newton's law of gravitation is given for two point masses m_1 and m_2 separated by a distance r
(a) The law can be applied as such to large bodies of masses m_1 and m_2 provided r is the distance between their centres of mass ☐
(b) The law cannot be applied to extended bodies ☐
(c) It is applicable only to regular geometrical figures such as cubes ☐

(*d*) It is applicable to uniform spherical distribution of mass, *r* being the distance between their centres. ☐

Q. 395. The gravitational potential inside a hollow spherical shell

(*a*) increases form zero at the centre to a maximum value at the surface ☐

(*b*) is zero throughout the interior ☐

(*c*) decreases from the centre ☐

(*d*) is uniform throughout the interior. ☐

Q. 396. If a tunnel could be dug through the earth from one side to the other along a diameter, and a particle is dropped into it

(*a*) the particle comes out at the other end of the tunnel ☐

(*b*) it falls down to the centre of the earth and remains there ☐

(*c*) it executes simple harmonic motion ☐

(*d*) it falls through with uniform acceleration.

Q. 397. The moon does not have an atmosphere like that of the earth because

(*a*) The sun's gravitational pull removes the gases ☐

(*b*) The earth's gravitational pull removes the gases ☐

(*c*) The escape velocity is low on the moon ☐

(*d*) The gravitational pull of the sun and the earth drains the moon of gases. ☐

Q. 398. The total energy of a satellite of mass *m* orbiting at a distance *R* from the centre of the earth (mass *M*) is

(*a*) $\frac{GMm}{R}$ ☐

(*b*) $\frac{GMm}{R^2}$ ☐

(*c*) $\frac{1}{2}\frac{GMm}{R}$ ☐

(*d*) $-\frac{1}{2}\frac{GMm}{R}$. ☐

Q. 399. A solid cylinder rolls down an inclined plane without slipping. An identical cylinder starting at the same instant slides down a frictionless. but otherwise identical inclined plane. As they reach the bottom

(*a*) the rolling cylinder has the greater speed □
(*b*) both have the same speed □
(*c*) the sliding cylinder has the greater speed □
(*d*) the sliding cylinder will come to rest. □

Q. 400. A sphere and a cylinder, having the same mass and radius, start from rest roll down the same inclined plane

(*a*) The sphere reaches the bottom first □
(*b*) The cylinder comes the bottom first □
(*c*) The cylinder reaches the bottom first □
(*d*) Both reach the bottom at the same time. □

Section V

Q. 401. A solid wooden sphere rolls down on two different inclined planes of the same height but different lengths. The speed as it reaches the bottom is

(*a*) greater on the shorter plane □
(*b*) smaller on the shorter plane □
(*c*) the same in each case □
(*d*) greater on the plane of greater incline. □

Q. 402. A diver leaving the diving board makes a somersault in the air

(*a*) This is due to gravitational force □
(*b*) His moment of inertia is decreased during the turn □
(*c*) His moment of inertia is increased □
(*d*) He pushes at the air for making the turn. □

Q. 403. The graph shown was obtained from experimental measurements of the period of oscillation T for different masses M placed in the scale pan on the lower end of the spring balance. The most likely reason for the line not passing through the origin is that the

(*a*) amplitude of the oscillations was too large □

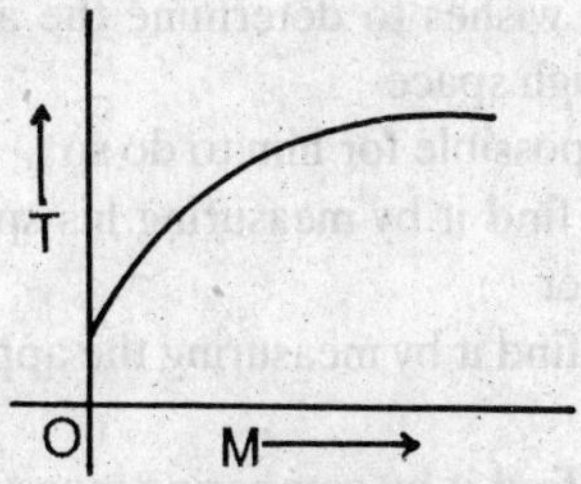

(*b*) spring did not obey Hooke's law ☐
(*c*) clock used needed regulating ☐
(*d*) mass of the pan was neglected. ☐

Q. 404. Which of the following relation is correct connecting elastic constant ?

(*a*) $\frac{9}{K} = \frac{3}{Y} + \frac{1}{9}$ ☐
(*b*) $\frac{9}{K} = \frac{3}{\eta} + \frac{1}{Y}$ ☐
(*c*) $\frac{9}{Y} = \frac{3}{\eta} + \frac{1}{K}$ ☐
(*d*) $\frac{9}{\eta} = \frac{3}{Y} + \frac{1}{K}$. ☐

Q. 405. Factor of safety may be defined by

(*a*) yielding stress/yielding stress ☐
(*b*) working stress/yielding stress ☐
(*c*) breaking stress/working stress ☐
(*d*) all of the above. ☐

Q. 406. The symbol c in the formula $E = mc^2$ represents

(*a*) the velocity of the particle ☐
(*b*) the charge of the particle ☐
(*c*) the velocity of the observer ☐
(*d*) the velocity of light. ☐

Q. 407. An observer in a closed laboratory

(*a*) can distinguish between an acceleration of the laboratory and the effects of a gravitational field ☐
(*b*) can measure the acceleration of the laboratory ☐
(*c*) can measure the velocity of the laboratory ☐
(*d*) can do none of the above. ☐

Q. 408. An observer wishes to determine the absolute speed of his motion through space

(*a*) it is impossible for him to do so ☐

(*b*) he can find it by measuring his speed relative to the other ☐

(*c*) he can find it by measuring the apparent speed of light ☐

(*d*) he can find it by comparing measurements made by another observer. ☐

Q. 409. According to the general theory of relatively, gravitation

(*a*) is really a general acceleration of the universe ☐

(*b*) acts only on moving bodies ☐

(*c*) is a distortion in space-time ☐

(*d*) is the same as in Newton's theory but is to be interpreted differently ☐

Q. 410. A crucial experiment that led to the theory of relativity was performed by

(*a*) Michelson and Morley ☐

(*b*) Pavel Cerenkov ☐

(*c*) Albert Einstein ☐

(*d*) Samuel celluer. ☐

Q. 411. When we measure the velocity and acceleration of an object, the figures we obtain

(*a*) are the same as those all other observers would find ☐

(*b*) refer to the motion of the object relative to the coordinates of four-dimensional space-time ☐

(*c*) refer only to the motion of the object relative to us ☐

(*d*) are always constant. ☐

Q. 412. An observer moves past a stationary electron on the earth's surface. His instruments measure

(*a*) only a magnetic field ☐

(*b*) only an electric field ☐

(*c*) both a magnetic and magnetic fields ☐

(*d*) neither electric nor magnetic field. ☐

Q. 413. If a ray of light is allowed to pass through a quartz crystal, then the two refracted rays obtained are

(*a*) plane polarised and plane of polarisation are parallel ☐
(*b*) circularly polarised in opposite direction ☐
(*c*) plane polarised and plane of polarisation are perpendicular ☐
(*d*) circularly polarised in same direction. ☐

Q. 414. Law of Malus states that

(*a*) when a ray of light passes through a quartz crystal it is split into two beams ☐
(*b*) there is an angle of polarisation for all reflected polarisation ☐
(*c*) the intensity of incident polarised ray is equal to the sum of intensities of the two reflected rays ☐
(*d*) the two reflected rays are plane polarised and their planes of polarisation are perpendicular. ☐

Q. 415. The failure of paraxial and marginal rays through the same point on reflection at a spherical mirror or refraction through a lens is known as

(*a*) chromatic aberration ☐
(*b*) spherical aberration ☐
(*c*) astigmatism ☐
(*d*) coma. ☐

Q. 416. Which of the following is used for the failure of a lens to form a sharp and distinct image of a white object ?

(*a*) Spherical aberration ☐
(*b*) Distortion ☐
(*c*) Chromatic aberration ☐
(*d*) Astigmatism. ☐

Q. 417. To reduce spherical aberration in optical instruments, which of the following should be used ?

(*a*) Plano-convex lenses ☐
(*b*) Concave lens ☐
(*c*) Spherical mirrors ☐

(*d*) Plane mirrors. ☐

Q. 418. Which of the following will not be there, if we use two lenses of the same material separated by a distance equal to mean focal lengths of the two lenses ?

(*a*) Spherical aberration ☐
(*b*) Chromatic aberration ☐
(*c*) Astigmatism ☐
(*d*) Distortion. ☐

Q. 419. Optically active substances are those which

(*a*) produce polarized light ☐
(*b*) rotate the plane of polarisation of polarized light ☐
(*c*) produce double refraction ☐
(*d*) converts a plane polarised light into circularly polarised light. ☐

Q. 420. We can speak of polarisation only in case of

(*a*) longitudinal wave motion ☐
(*b*) transverse waves ☐
(*c*) Both the above ☐
(*d*) None of these. ☐

Q. 421. One of the devices to produce plane polarised light is

(*a*) a prism ☐
(*b*) a bi-prism ☐
(*c*) a Nicol prism ☐
(*d*) None of these. ☐

Q. 422. A polaroid is

(*a*) a red light filter ☐
(*b*) a device used for analysing polarised light ☐
(*c*) a device used in polarimeter ☐
(*d*) an adjustable shutter. ☐

Q. 423. The spectrum of light from the sun is found to be crossed by a large number of fine lines called

(*a*) Fresnel lines ☐
(*b*) Resonance levels ☐
(*c*) Fraunhoffer lines ☐
(*d*) W. Herschels' lines. ☐

Q. 424. Bells are made of metals and not of wood, because

(*a*) the thermal conductivity of metal is greater than that of wood ☐

(*b*) the density of metal is greater than that of wood ☐

(*c*) the sound is not conducted by metals but is radiated ☐

(*d*) wood dampens vibrations while metals are elastic. ☐

Q. 425. The combi ation of two or more notes which gives you melodious effect on ear is known is

(*a*) concord ☐

(*b*) chords ☐

(*c*) beats ☐

(*d*) overtones. ☐

Q. 426. Beats occur because of

(*a*) reflection ☐

(*b*) refraction ☐

(*c*) interference ☐

(*d*) Doppler effect. ☐

Q. 427. Which of the following is the practical application of reflection of sound ?

(*a*) Ear trumpet ☐

(*b*) Speaking tube ☐

(*c*) Sound board ☐

(*d*) All of the above. ☐

Q. 428. Which of the following is the equation of a simple harmonic motion ?

(*a*) $x = a \sin wt$ ☐

(*b*) $x = a \cos wt$ ☐

(*c*) $x = a (\sin^2 wt + \cos^2 wt)$ ☐

(*d*) None of the above. ☐

Q. 429. When the two sound waves of the same frequency and amplitude are 100 degrees out of phase, the result is

(*a*) beats ☐

(*b*) increased loudness ☐

(*c*) resonance ☐

(*d*) silence. ☐

Q. 430. When a source of sound is in motion towards a stationary observer, the effect observed is

(*a*) increases in velocity of sound only ☐
(*b*) decrease in velocity of sound only ☐
(*c*) increase in frequency of sound only ☐
(*d*) increase in velocity as well as frequency of sound. ☐

Q. 431. The distance from crest to crest of any wave is called its

(*a*) frequency ☐
(*b*) wavelength ☐
(*c*) speed ☐
(*d*) amplitude. ☐

Q. 432. A sound whose frequency is 50 Hz

(*a*) has a relatively short wavelength ☐
(*b*) has a relatively long wavelength ☐
(*c*) is very loud ☐
(*d*) can travel through vacuum. ☐

Q. 433. Which of the following is a sensory effect rather than a measurable physical quantity ?

(*a*) Frequency ☐
(*b*) Intensity ☐
(*c*) Quality ☐
(*d*) Waveform. ☐

Q. 434. In a fire siren, a jet of air is directed again a series of evenly-spaced holes in a rotating disc. As the disc speeds up, the tone

(*a*) increases in frequency ☐
(*b*) increases in wavelength ☐
(*c*) drops in pitch ☐
(*d*) maintains a constant pitch. ☐

Q. 435. An organ pipe vibrates in fundamental resonance with its medium, first as air, then hydrogen and then methane. The velocity of sound in the three media is different, owing to which in the three cases

(*a*) the wavelength changes ☐
(*b*) the frequency changes ☐
(*c*) the wavelength and frequency both changes ☐

(*d*) the wavelength and frequency both remain invariant. ☐

Q. 436. Which of the following instruments is used to record sound ?

(*a*) Amplifier ☐
(*b*) Headphone ☐
(*c*) Dictaphone ☐
(*d*) Phonograph. ☐

Q. 437. If T is the tension applied to the string and m is the mass per unit length of the string, then the velocity of a transverse wave along a stretched string is given by

(*a*) $\sqrt{T/m}$ ☐
(*b*) $\sqrt{m/T}$ ☐
(*c*) $\sqrt{m \times T}$ ☐
(*d*) $m\sqrt{T}$. ☐

Q. 438. Sound will not travel through a

(*a*) solid ☐
(*b*) liquid ☐
(*c*) gas ☐
(*d*) vacuum. ☐

Q. 439. A body in equilibrium may not have

(*a*) momentum ☐
(*b*) velocity ☐
(*c*) acceleration ☐
(*d*) kinetic energy. ☐

Q. 440. If a car is to gain momentum it must

(*a*) lose inertia ☐
(*b*) accelerate ☐
(*c*) move rapidly ☐
(*d*) lose weight. ☐

Q. 441. A bomb dropped from an aeroplane explodes in air, its total

(*a*) momentum decreases ☐
(*b*) momentum increases ☐
(*c*) kinetic energy increases ☐
(*d*) kinetic energy decreases. ☐

Q. 442. If the polar ice caps melts, the length of the day will

(*a*) increase
(*b*) decrease
(*c*) remain the same
(*d*) insufficient information to predict.

Q. 443. Eight steel balls are suspended from equal heights by threads so that they are all in contact. If two balls are pulled away to the left and released
(*a*) one ball will leave form the right
(*b*) two balls will leave from the right at approximately the same speed
(*c*) either (*a*) or (*b*)might apply with twice the speed
(*d*) neither (*a*) nor (*b*) applies.

Q. 444. Swimming is possible because of which law of motion ?
(*a*) Second
(*b*) First
(*c*) Third
(*d*) None of these.

Q. 445. The total angular momentum of a system of particles
(*a*) changes when a net external force acts upon the system
(*b*) remains constant under all circumstances
(*c*) changes when a net external torque acts upon the system
(*d*) may or may not change under the influence of a net external torque depending upon the direction of the torque.

Q. 446. The current in the heater circuit of a vacuum tube is 1.5 amperes in section A.C. and 1.7 amperes in section B.D., then the magnitude of plate current is
(*a*) $0.2A$
(*b*) $0.1A$
(*c*) $0.6A$.
(*d*) $3.2A$.

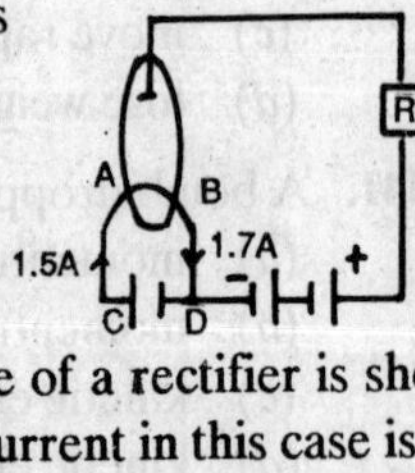

Q. 447. The output current versus time curve of a rectifier is shown below. The average value of output current in this case is
(*a*) 0

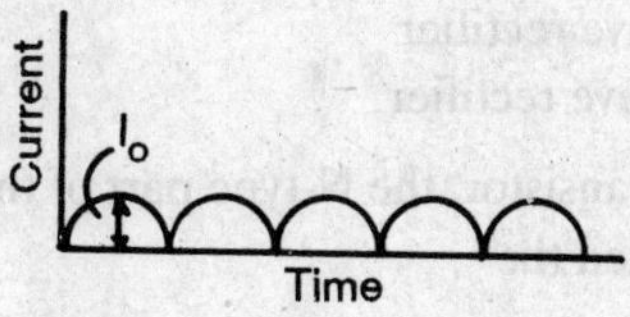

(b) $\frac{I_0}{2}$ ☐

(c) $\frac{2I_0}{\pi}$ ☐

(d) I_o. ☐

Q. 448. The first thermionic valve which was designed by Fleming in 1904 was known as

(a) rectifier ☐
(b) transistor ☐
(c) Fleming valve ☐
(d) None of these. ☐

Q. 449. Since a diode permits the flow of current only in one direction so it can be used as

(a) an oscillator ☐
(b) a rectifier ☐
(c) a photo detector ☐
(d) None of these. ☐

Q. 450. Transistors are made from

(a) plastics ☐
(b) metals ☐
(c) insulators ☐
(d) doped semi-conductors. ☐

Q. 451. A triode valve will work as a better amplifier if

(a) $\Delta p/\Delta V_g$ is very small ☐
(b) $\Delta v/\Delta I_p$ is small ☐
(c) $\Delta v/\Delta I_p$ is large ☐
(d) $\Delta p/\Delta V_g$ is large. ☐

Q. 452. Which of the following is not an application of diodes ?

(a) A filter ☐
(b) Bridge rectifier ☐

(*c*) Full-wave rectifier ☐
(*d*) Half-wave rectifier. ☐

Q. 452. In an NPN-transistor, the N-type part of the forward- biased section is called the
(*a*) emitter ☐
(*b*) collector ☐
(*c*) base ☐
(*d*) donor. ☐

Q. 454. In Young's experiment interference of light with two slits, reinforcement occurs whensin $\theta = \frac{n\lambda}{d}$, d is
(*a*) distance from slits to screen ☐
(*b*) distance between dark and bright fringes ☐
(*c*) distance between slits ☐
(*d*) width of the *n*th fringe. ☐

Q. 455. The path difference corresponding to a phase difference π is
(*a*) λ ☐
(*b*) $\frac{\lambda}{2}$ ☐
(*c*) 2λ ☐
(*d*) $\frac{\lambda}{4}$. ☐

Q. 456. Oil floating on water looks coloured due to the interference of light ; what should be the order of magnitude of the thickness of oil layer in order that this effect may be observed
(*a*) 10 A ☐
(*b*) 100 A ☐
(*c*) 10,000 A ☐
(*d*) 1 cm. ☐

Q. 457. In the absence of atmosphere the duration of day on earth will be
(*a*) shorter ☐
(*b*) longer ☐
(*c*) the same ☐
(*d*) longer or shorter depending on the weather. ☐

Q. 458. A laser beam may be used to measure very large distance because, it is

(*a*) unidirectional ☐
(*b*) monochromatic ☐
(*c*) coherent ☐
(*d*) not absorbed. ☐

Q. 459. In a compound microscope, magnification will be large if the focal length of eye-piece is

(*a*) large ☐
(*b*) small ☐
(*c*) equal to that of the objective ☐
(*d*) None of these. ☐

Q. 460. A fly is found to be sitting on a telescope when it is focussed towards the moon, what effect expected on the photograph of moon ?

(*a*) Intensity remains unchanged ☐
(*b*) No effect ☐
(*c*) The complete field of view is blocked ☐
(*d*) Coloured image will be seen. ☐

Q. 461. In Young's double slit experiment the path difference between two waves producing constructive interference is equal to

(*a*) $n\lambda$ ☐
(*b*) $(2n+1)\lambda/2$ ☐
(*c*) both the above ☐
(*d*) None of these. ☐

Q. 462. The fringes seen in Young's double slit experiment are an evidence of

(*a*) corpuscular nature of light ☐
(*b*) wave nature of light ☐
(*c*) dual nature of light ☐
(*d*) None of these. ☐

Q. 463. Chromatic aberration in the formation of images by a lens arises because

(*a*) of non-paraxial rays ☐
(*b*) radii of curvature of the two sides are not same ☐
(*c*) of the defect is grinding ☐
(*d*) the focal length varies with wave-length. ☐

Q. 464. In a Young's interference experiment an interference pattern of bright and dark fringes is formed on screen (see Fig. below) by light of wavelength λ coming from two narrow slits X and Y. The difference between PX and PY is

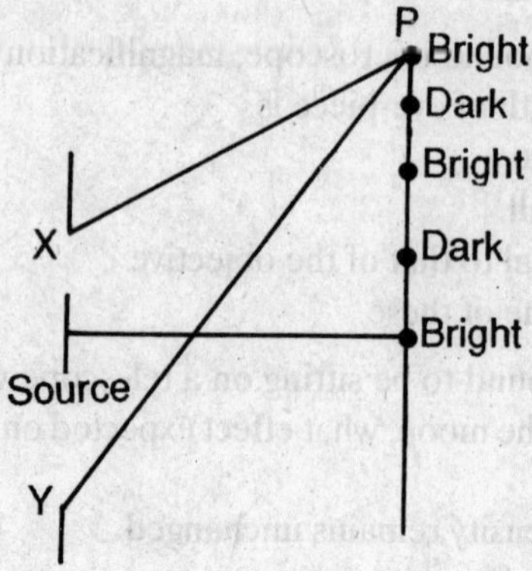

(*a*) λ ☐
(*b*) $\frac{3\lambda}{2}$ ☐
(*c*) 2λ ☐
(*d*) $\frac{5\lambda}{2}$. ☐

Q. 465. The soap bubbles of unequal sizes are blown at the ends of *T*-tubes. When the bubbles are put in communication with each other

(*a*) air presses from smaller to bigger ☐
(*b*) air presses from bigger to smaller bubble ☐
(*c*) there is no change of size ☐
(*d*) both the bubbles collapse. ☐

Q. 466. A liquid rises to a height *h* in a capillary tube. If the height of the capillary tube above the liquid surface is slightly less than *h*

(*a*) there will be no capillary rise ☐
(*b*) there is no meniscus ☐
(*c*) the radius of curvature of the meniscus increase ☐
(*d*) the radius of curvature of the meniscus decreases. ☐

Q. 467. Surface tension of liquid is lowered
(*a*) by any impurity in it
(*b*) by a highly soluble impurity
(*c*) by a feebly soluble impurity
(*d*) by inorganic salts only.

Q. 468. If T_o is the surface tension of water at O° C, the surface tension T at a slightly higher temperature t°C is given by
(*a*) $T = T_o(1+xt^2)$
(*b*) $T = T_o(1+xt)$
(*c*) $T = T_o(1-xt)$
(*d*) $T = T_o(1-xt^2)$.

Q. 469. Phospher bronze is often used for torsional suspension because
(*a*) its rigidity modulus is high
(*b*) its elastic limit is high
(*c*) it has no elastic after effect
(*d*) it has elastic hysteresis.

Q. 470. A simple shear is equivalent at right angles
(*a*) two equal extensions at right angles
(*b*) an extension and an equal compression at right angles
(*c*) two equal compressions at right angles
(*d*) an extension and a compression at 45°.

Q. 471. Bernoulli's equation is applicable to points
(*a*) in a steadily flowing liquid
(*b*) in a stream line
(*c*) in a straight line perpendicular to stream line
(*d*) in any non-viscous liquid.

Q. 472. The venturimeter is an instrument for measuring
(*a*) the viscosity of a liquid
(*b*) the flow speed of a liquid
(*c*) the compressibility of a fluid
(*d*) the specific gravity of the liquid.

Q. 473. The airplane lift is based on
(*a*) Archimedes principle
(*b*) Conservation of linear momentum
(*c*) Conservation of energy

(*d*) Bernoulli's principle. ☐

Q. 474. The velocity of efflux of a liquid from an orifice is

(*a*) $\sqrt{gh}$ ☐

(*b*) $\sqrt{2hg}$ ☐

(*c*) $\frac{gh}{2}$ ☐

(*d*) $2\sqrt{gh}$. ☐

where *h* is depth of the orrific below the surface.

Q. 475. The SI unit of coefficient of viscosity is

(*a*) newton m/s ☐

(*b*) newton second meter ☐

(*c*) newton second/meter^2 ☐

(*d*) newton meter/second^2. ☐

Q. 476. In turbulent motion of a viscous liquid,

(*a*) the rate of flow is inversely proportional to viscosity ☐

(*b*) the rate of flow is directly proportional to viscosity ☐

(*c*) the rate of flow is inversely proportional to the square of the viscosity ☐

(*d*) the rate of flow is independent of viscosity. ☐

Q. 477. The surface of liquid is is some what similar to a stretched membrane because

(*a*) there is an elastic stress is the surface ☐

(*b*) the tension in surface increase if the area is increased ☐

(*c*) the surface has a natural tendency to contract ☐

(*d*) ripples can be produced on the surface. ☐

Q. 478. Cosmic rays are

(*a*) *X*-rays coming from atmosphere ☐

(*b*) γ-rays coming from sun ☐

(*c*) u.v-rays coming from space ☐

(*d*) high energy charged particles coming from outer space. ☐

Q. 479. If the half-life ($T_{1/2}$) of a radioactive element is one year. In how much will the sample of such an element reduce to 1/8th of its mass ?

(*a*) 1/2 year
(*b*) 2 years
(*c*) 3 years
(*d*) 4 years.

Q. 480. When a positron is emitted, it cause
(*a*) mass number to increase by one
(*b*) mass number to decrease by one
(*c*) atomic number to increase by one
(*d*) atomic number tc decrease by one.

Q. 481. In the decay of a radioactive element, $N = N_o\, e^{-\lambda . t}$, here λ stands for
(*a*) wavelength
(*b*) life-time
(*c*) decay constant
(*d*) None of these.

Q. 482. What for are cadmium or boron rods used in a nuclear reactor ?
(*a*) To slow down the reaction
(*b*) To absorb the excess the excess neutrons
(*c*) To absorb α-particles
(*d*) None of these.

Q. 483. To sustain a chain reaction, a mass of *U*-235 must be
(*a*) at very high temperature
(*b*) bombarded with neutrons
(*c*) exposed to γ-rays
(*d*) None of these.

Q. 484. In the following reaction, *P* stands for what ?
$zX^a + {}_0n^1 \rightarrow {}_{z+1}Y^{a+1} + P$
(*a*) Electron
(*b*) Neutron
(*c*) Proton
(*d*) None of these.

Q. 485. The reason of neutrons being more effective as bombarding particles as compared to protons is that
(*a*) neutrons have more kinetic energy
(*b*) neutron emit β-particles

(c) neutron remain unaffected by electric fields ☐
(d) neutrons exerts stronger nuclear forces. ☐

Q. 486. Every point on any wave front can be considered as a new source of waves. This statement is known as
(a) Huygen's wave theory ☐
(b) Michelson's wave theory ☐
(c) Raman's effect ☐
(d) Rayleih's wave theory. ☐

Q. 487. The failure of the paraxial and marginal rays through the same point on reflection at a spherical mirror or refraction of light through a lens is called
(a) polarisation ☐
(b) interfernence ☐
(c) spherical abberration ☐
(d) diffraction. ☐

Q. 488. Which one of the following conditions a good microscope objective should satisfy ?
(a) It should be free from chromatic aberration ☐
(b) It should be free from spherical aberration ☐
(c) A good deal of light should enter it because the object to the viewed is very small ☐
(d) All of the above. ☐

Q. 489. A *wave front* is an imaginary surface where
(a) phase changes with constant rate in all directions along the surface ☐
(b) phase changes with rate of $2\pi/\lambda$ per unit length in all directions along the surface ☐
(c) constant phase difference is always maintained ☐
(d) phase is always same for all points. ☐

Q. 490. If the whole of Young's experiment is performed in water, which of the following changes will occur ?
(a) The fringe width will increase ☐
(b) The fringe width will decrease ☐
(c) Fringe width will remain unchanged ☐
(d) No fringe will be seen. ☐

Q. 491. Which is the most essential condition for producting interference ?

(*a*) The waves emitted by the two sources should either have the same phase or a constant difference in phase ☐

(*b*) The two sources must be very close to each other ☐

(*c*) The two sources be very narrow ☐

(*d*) All of the above. ☐

Q. 492. What is a "Zoom lens " ?

(*a*) It is a lens having fixed focal length ☐

(*b*) It is a lens having variable focal length ☐

(*c*) It is a lens used in radii telescope ☐

(*d*) None of the above. ☐

Q. 493. Longitudinal waves *cannot* be

(*a*) reflected ☐

(*b*) diffracted ☐

(*c*) dispersed ☐

(*d*) polarised. ☐

Q. 494. Images formed from real objects by diverging lenses are always

(*a*) virtual ☐

(*b*) enlarged ☐

(*c*) inverted ☐

(*d*) real. ☐

Q. 495. A diverging lens may not have

(*a*) a negative focal length ☐

(*b*) a positive focal length ☐

(*c*) one plane surface ☐

(*d*) one convex surface. ☐

Q. 496. For a compound microscope, the image formed by the eye-piece is

(*a*) real ☐

(*b*) inverted ☐

(*c*) erect ☐

(*d*) diminished. ☐

Q. 497. For optical phenomenon, as described by electromagnetic theory, it is the electric vector which is involved the magnetic vector introduction in theory is

(*a*) only customary ☐
(*b*) for representing the field energy or density flux ☐
(*c*) because with an electric vector association ☐
(*d*) for explaining anamalous dispersion. ☐

Q. 498. A constant voltage is applied between the two ends of a uniform metallic wire. Some heat is produced in it. The heat doubles if

(*a*) both length and the radius of the wire are halved ☐
(*b*) both length and the radius of the wire are doubled ☐
(*c*) the radius of the wire gets doubled ☐
(*d*) the length of the wire gets doubled. ☐

Q. 499. A rectangular loop carrying a current *i* is situated near a long straight wire such that the wire is parallel to one of the sides of the loop and is in the plane of the loop. If the steady current *I* is established in the wire as shown in figure, the loop will

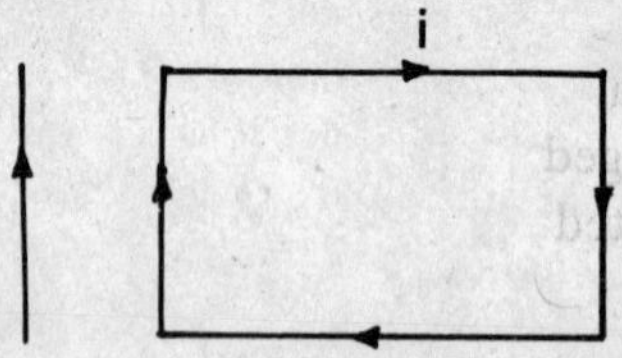

(*a*) rotate about an axis parallel to the wire ☐
(*b*) move away from the wire ☐
(*c*) move towards the wire ☐
(*d*) remain stationary. ☐

Q. 500. Which of the following instruments is designed to change electrical energy into sound energy ?

(*a*) Amplifier ☐
(*b*) Generator ☐
(*c*) Loudspeaker ☐
(*d*) Microphone. ☐

Section VI

Q. 501. Which voltmeter and ammeter should be used to determine as accurately as possible a resistance of approximately 200 Ω ?

	Resistance of voltmeter	*Resistance of ammeter*	
(*a*)	5000Ω	0.2Ω	☐
(*b*)	5000Ω	2Ω	☐
(*c*)	200Ω	200Ω	☐
(*d*)	2Ω	1000Ω.	☐

Q. 502. Which one of the following is not an advantage of AC over DC ?

(*a*) The AC shows skin effect ☐
(*b*) In AC a wide range of voltages is possible ☐
(*c*) Lines losses in AC transmission lines are negligible ☐
(*d*) In AC the magnitude of current can be reduced by a choke. ☐

Q. 503. You are sitting in a room in which uniform magnetic field is present in vertically downward direction. When an electron is projected in horizontal direction it will be moving in circular path with constant speed

(*a*) clockwise in vertical plane ☐
(*b*) clockwise in horizontal plane ☐
(*c*) anti-clockwise in horizontal plane ☐
(*d*) anti-clockwise in vertical plane. ☐

Q. 504. Wattless current is said to flow when phase angle between virtual current and virtual voltage is

(*a*) 0° ☐
(*b*) 90° ☐
(*c*) 180° ☐
(*d*) None of these. ☐

Q. 505. Slip rings are used in

(*a*) DC dynamo ☐

(b) AC dynamo
(c) don't know
(d) None of these is correct.

Q. 506. Split-ring commutor is typical of
(a) a D.C. generator
(b) an A.C. generator
(c) a transformer
(d) a rectifier.

Q. 507. The frequency of A.C. is
(a) the number of cycles of the current per minute
(b) the number of cycles per second
(c) Both of these
(d) None of these.

Q. 508. An α-particle of energy 5 MeV is scatterred through 180° by a fixed uranium nucleus. The distance of closest approach is of the order of
(a) 1 Å
(b) 10^{-10} cm
(c) 10^{-12} cm
(d) 10^{-15} cm.

Q. 509. If the work function of the metal is ϕ and the frequency of incident light is v, there is not emission of photoelectrons when
(a) $v < \frac{\phi}{h}$
(b) $v = \frac{\phi}{h}$
(c) $v > \frac{\phi}{h}$
(d) $v \geq \frac{\phi}{h}$.

Q. 510. Hydrogen emits blue light when it changes from $n = 4$ energy level to $n = 2$ energy level. Which colour of light will be emitted when the electron in this atom jumps from $n = 5$ to $n = 2$?
(a) Red
(b) Yellow
(c) Green

(*d*) Violet. ☐

Q. 511. We may state that

(*i*) The energy E of a photon of frequency v, is $E = hv$, where h is plank's constant.

(*ii*) The momentum p of a photon is $p = \frac{h}{\lambda}$ where λ is the wave length of the photon.

From the above statement we may conclude that the wave velocity of light is equal to

(*a*) 3×10^{10} cm/sec ☐
(*b*) the ratio of $\frac{E}{p}$ ☐
(*c*) the product $E \times p$ ☐
(*d*) None of these. ☐

Q. 512. In the following reaction radiations emitted are in the sequence

$$ {}^{A}_{Z}X \longrightarrow {}^{A}_{Z^{+1}}Y \longrightarrow {}^{A-1}_{Z^{-1}}K \longrightarrow {}^{A-4}_{Z^{-1}}K $$

(*a*) α, β, γ ☐
(*b*) β, α, γ ☐
(*c*) γ, α, β ☐
(*d*) β, γ, α. ☐

Q. 513. The radius (R) of a nucleus charges with the nucleon number (A) by the nucleus as

(*a*) $R = 1{\cdot}3 \times 10^{-15} A^{2/3}$ metre ☐
(*b*) $R = 1{\cdot}3 \times 10^{-15} A^{1/3}$ metre ☐
(*c*) $R = 1{\cdot}3 \times 10^{-15} A^{8/3}$ meter ☐
(*d*) None of these. ☐

Q. 514. When do two protons attract each other ?

(*a*) The distance between them is 10^{-10} m ☐
(*b*) The distance between them is 10^{-1} m ☐
(*c*) The distance between them is 10^{-15} m ☐
(*d*) This is never possible. ☐

Q. 515. When two nuclei of masses m_1 and m_2 fuse to form nucleus of mass m and some energy is released then

(*a*) $(m_1 + m_2) > m$ ☐

(b) $(m_1 + m_2) < m$
(c) $m_1 + m_2 = m$
(d) $m_1 - m_2 = m$.

Q. 516. Huygen's concept of secondary waves
(a) allows us to find the focal length of a thick lens
(b) is a geometrical method to find a wave front
(c) is used to determine the velocity of light
(d) is used to explain polarisation.

Q. 517. Which of the following scientists measured the speed of the light by using an eight-sided rotating mirror ?
(a) Fizeau
(b) Roemer
(c) Galileo
(d) Michelson.

Q. 518. Which of the following terms best describes the nature of light from the modern point of view ?
(a) Photons
(b) Waves
(c) Particles
(d) Rays.

Q. 519. Which of the following light sources of equal radiant power appears the brightest ?
(a) Red light
(b) Green light
(c) Blue light
(d) All are the same.

Q. 520. During a solar eclipse, a person on earth with his eyes protected observes that part of the sun is darkened but never all of it. This person lies in the
(a) umbra
(b) penumbra
(c) both (a) and (b)
(d) Neither (a) nor (b).

Q. 521. In the construction of Huygen's eye-piece, the focal length of the field and the eye lens should be
(a) 3 : 1

(b) 1 : 3

(c) 1 : 2

(d) 1 : 1.

Q. 522. Laser light is considered to be coherent because it consists of

(a) many wavelengths

(b) uncoordinated wavelengths

(c) coordinated waves of exactly the same wavelength

(d) divergent beams.

Q. 523. In astronomical telescope the magnification is given by (for normal eye)

(a) $\frac{f_o}{f_e}$

(b) D

(c) $1 + \frac{D}{f_e}$

(d) $\frac{D}{f_e}$.

Q. 524. Which of the following is used to produce plane polarised light ?

(a) A nicol prism

(b) A crystal

(c) A biprism

(d) A half-wave plate.

Q. 525. When a monochromatic wave enters from one medium into another, the property that remains unchanged is

(a) amplitude

(b) velocity

(c) frequency

(d) None of these

Q. 526. A diffration pattern is obtained using a beam of red light. What happens if the red light is replaced by blue light ?

(a) No change

(b) Band disappears

(c) Diffraction bands become narrower and crowded together

(d) Bands become broader and farther apart.

Q. 527. The Great Indian scientist Prof. C.V. Raman was awarded a noble prize for his work that was concerned with

(*a*) nuclear energy ☐
(*b*) laser ☐
(*c*) scattering of light ☐
(*d*) transistors. ☐

Q. 528. A hydrometer floats to a particular level in sea water. In fresh water it

(*a*) floats lower ☐
(*b*) sinks completely ☐
(*c*) floats higher ☐
(*d*) floats at the same level. ☐

Q. 529. A weight is suspended from the middle of a rope whose ends are at the same level. In order for the rope to be perfectly horizontal, the forces applied to the ends of the rope

(*a*) must be equal to the weight ☐
(*b*) must be greater than the weight ☐
(*c*) might be so great as to break the rope ☐
(*d*) must be infinite. ☐

Q. 530. It is desired to reach a point just opposite on the other bank of the river. Then one should row the boat

(*a*) up-stream ☐
(*b*) inclined to the bank up stream ☐
(*c*) down stream ☐
(*d*) inclined to the bank down-stream. ☐

Q. 531. A thin film of liquid is enclosed between two glass plates. It is difficult to separate the plates on account of

(*a*) surface tension ☐
(*b*) atmospheric pressure ☐
(*c*) viscosity ☐
(*d*) friction. ☐

Q. 532. The working of a rocket is based on the principle of

(*a*) electricity ☐
(*b*) conservation of momentum ☐
(*c*) Kepler's law ☐
(*d*) Newton's law. ☐

Q. 533. A hunter aims and shoots an arrow at a monkey on a pole. Monkey moves horizontally just before reaching the arrow to it but even then arrow strikes the monkey. It shows that velocity of monkey is equal to

(*a*) vertical component of the velocity of the arrow ☐
(*b*) horizontal component of the velocity of the arrow ☐
(*c*) original velocity of the arrow ☐
(*d*) none of the above statements. ☐

Q. 534. The action and reaction forces referred to in Newton's third law or motion

(*a*) must act upon the same body ☐
(*b*) may act upon different bodies ☐
(*c*) need not to be equal in magnitude but must have the same line of action ☐
(*d*) must be equal in magnitude but need not have the same line of action. ☐

Q. 535. An automobile that is towing a trailer is accelerating on a level road. The force that the automobile exerts on the trailer is

(*a*) equal to the force the trailer exerts on the automobile ☐
(*b*) greater than the force the trailer exerts on the automobile ☐
(*c*) equal to the force the trailer exerts on the road ☐
(*d*) equal to the force the road exerts on the trailer. ☐

Q. 536. The time period of simple pendulum can be doubled by :

(*a*) increasing the length of the pendulum 2 times ☐
(*b*) increasing the length of the pendulum 4 times ☐
(*c*) increasing the mass of the bob 4 times ☐
(*d*) increasing the mass of bob 2 times. ☐

Q. 537. The static friction is

(*a*) always greater than the dynamic friction ☐
(*b*) always less than the dynamic friction ☐
(*c*) equal to the dynamic friction
(*d*) sometimes greater and sometimes less than the dynamic friction. ☐

Q. 538. If you float in water with just your nose out, the average density of your body must be :

(*a*) same as that of the water ☐
(*b*) greater than that of water ☐
(*c*) less than that of water ☐
(*d*) None of the above said. ☐

Q. 539. A piece of weighed wood just floats in water. When placed in alcohol (of lower denisty) it will

(*a*) float lower ☐
(*b*) float higher ☐
(*c*) stay as before ☐
(*d*) sink. ☐

Q. 540. Who discovered photoelectric effect ?

(*a*) Lenard ☐
(*b*) Einstein ☐
(*c*) Hallwachs ☐
(*d*) Max Planck. ☐

Q. 541. The strength of photoelectric current depends upon

(*a*) intensity of incident light ☐
(*b*) frequency of incident light ☐
(*c*) angle of incident light ☐
(*d*) distance between anode and cathode. ☐

Q. 542. The stopping potential for the photoelectron from photocell is

(*a*) directly proportional to intensity of incident light ☐
(*b*) inversely proportional to intensity of incident light ☐
(*c*) directly proportional to frequency of incident light ☐
(*d*) inversely proportional to frequency of incident light. ☐

Q. 543. The study of photoelectric cell is useful in understanding of

(*a*) Bohr's atomic model ☐
(*b*) Compton effect ☐
(*c*) Quantum nature of light ☐
(*d*) None of these. ☐

Q. 544. Energy released in the process of fusion is
(*a*) equal to fission ☐
(*b*) always greater than fission ☐
(*c*) always less than fission ☐
(*d*) some times greater than and some times less than fission. ☐

Q. 545. (1) Fission is a thermonuclear process.
(2) Fussion is a thermonuclear process.
(3) Fusion is exothermic.
(4) Fission is exothermic.
Choose the correct answer :
(*a*) 1 and 2 are correct ☐
(*b*) 2 and 3 are correct ☐
(*c*) 1 and 4 are correct ☐
(*d*) 2 and 4 are correct. ☐

Q. 546. The critical mass of fissionable uranium 235 can be reduced by
(*a*) adding impurities to it ☐
(*b*) heating material ☐
(*c*) surrounding it by neutron reflecting material ☐
(*d*) surrounding it by neutron absorbing material. ☐

Q. 547. If the multiplication factor is greater than unity in a nuclear fission reaction involving neutron capture, then
(*a*) the process will stop after some time ☐
(*b*) the reaction will be self-sustaining with no chance of explosion ☐
(*c*) an explosion may result ☐
(*d*) an explosion cannot occur. ☐

Q. 548. The element not occurring in nature is
(*a*) ${}_{92}U^{233}$ ☐
(*b*) ${}_{92}U^{235}$ ☐
(*c*) ${}_{92}U^{238}$ ☐
(*d*) ${}_{92}U^{232}$. ☐

Q. 549. Enriched uranium is ordinary urnium-238
(*a*) whose density has been increased by the application of pressure ☐

(b) in which gold has been mixed under pressure ☐
(c) in which proportion of the rare isotope *U*-235 has been raised beyond 0.7% by adding to it additional amount of *U*-235 ☐
(d) in which some thorium-232 has been mixed under pressure. ☐

Q. 550. A slow neutron can cause fission in
(a) uranium-238 ☐
(b) uranium-235 ☐
(c) hydrogen-1 ☐
(d) thorium-232. ☐

Q. 551. According to the modern theory of relativity, Newtonian mechanics is
(a) totally incorrect ☐
(b) approximately correct for any velocity ☐
(c) approximately correct for speeds much less than the speed of light ☐
(d) correct for all velocities. ☐

Q. 552. A clock of mass m and length L records a time interval Δt. If this clock moves with a velocity of 0.6*C*, which of the following statement is *not* true ?
(a) The mass will be larger ☐
(b) The length will shorter ☐
(c) The time interval will be longer ☐
(d) The time interval will be shorter. ☐

Q. 553. Work funcion is generally expressed in
(a) electron volt ☐
(b) joule ☐
(c) newton ☐
(d) gauss. ☐

Q. 554. Who discovered the thermionic emission ?
(a) Fleming ☐
(b) Lee de Forest ☐
(c) Gamows ☐
(d) Rutherford. ☐

Q. 555. Wheel with angular momentum of 10 kg. m^2/s has a moment of intertia equal to 6.5 kg m^2. Its angular speed is

(*a*) 5 rad/s ☐
(*b*) 20 rad/s ☐
(*c*) 40 rad/s ☐
(*d*) 0.02 rad/s. ☐

Q. 556. An object swings the end of a string in uniform circular motion. Which of the following changes would not cause an increased centripetal force ?-

(*a*) a longer string ☐
(*b*) a greater linear speed ☐
(*c*) a shorter string ☐
(*d*) a larger mass. ☐

Q. 557. In a rigid body undergoing uniform circular motion, a particle that is a distance *R* from the axis of rotation

(*a*) has a angular velocity inversely proportional to *R* ☐
(*b*) has an angular velocity proportional to *R* ☐
(*c*) has a linear speed proportional to *R* ☐
(*d*) has a linear speed inversely proportional to *R*. ☐

Q. 558. The unit for the moment of inertia in the M.K.S. system is

(*a*) kg m^3 ☐
(*b*) kg m^2 ☐
(*c*) kg/m^2 ☐
(*d*) kg/m. ☐

Q. 559. An object is travelling in a circle with a constant speed. Its acceleration is constant in

(*a*) magnitude only ☐
(*b*) both magnitude and direction ☐
(*c*) direction only ☐
(*d*) neither magnitude nor direction. ☐

Q. 560. Which of the following formulae for the moment of inertia (M.I.) of some simple cases is not correct ?

(*a*) The M.I. of a uniform rod of mass *M* and length *l* about an axis through its centre and perpendicular to its length is $I = Ml^2/12$ ☐
(*b*) The M.I. of a uniform circular disc of Mass *M* and radius *r* about an axis through its centre and

perpendicular to the plane of the disc is $I = 1/2\, Mr^2$ ☐

(c) The M.I. in case of (b) about the axis through one end and perpendicular to its length is $I = Ml^2/4$ ☐

(d) The M.I. of a sphere of mass M and radius r about any diameter is $I = 2/5Mr^2$. ☐

Q. 561. The acceleration of a body undergoing uniform circular motion is constant in

(a) magnitude only ☐
(b) both magnitude and direction ☐
(c) direction only ☐
(d) neither magnitude nor direction. ☐

Q. 562. A quantity not directly involved in the rotational motion body is

(a) mass ☐
(b) torque ☐
(c) moment of inertia ☐
(d) angular velocity. ☐

Q. 563. When a space vehicle takes off from the ground to orbit the earth as a satellite, the velocity that it must acquire to circle the earth is called

(a) Escape velocity ☐
(b) Orbiting velocity ☐
(c) Constant velocity ☐
(d) None of the above. ☐

Q. 564. The escape velocity from the earth's surface is

(a) 1.7 km/sec. ☐
(b) 1.2 km/sec. ☐
(c) 11.2 km/sec. ☐
(d) 10.2 km/sec. ☐

Q. 565. The escape velocity

(a) increase with the increase of the mass of the body ☐
(b) depends on the type of body used ☐
(c) is independent of mass of the body ☐
(d) decreases with the increase of the mass of the body. ☐

Q. 566. A glass jar is fastented on a horizontal turn-table near its rim and a candle is placed inside the jar. When the candle is lit, the flame will point in a vertical direction owing to the upward motion of the hot gases formed in the combustion process. If the turn table is set into rotation, the flame will be deflected:

(*a*) away from center of circle ☐

(*b*) along the tangent to circle ☐

(*c*) towards the center of circle ☐

(*d*) remains as it was. ☐

Q. 567. If M is the mass of the earth and R its radius, the velocity of escape from earth's surface is given by

(*a*) $\frac{2GM}{R}$ ☐

(*b*) $\frac{GM^2}{3}$ ☐

(*c*) $\sqrt{2gR}$ ☐

(*d*) $\frac{3gR}{2}$. ☐

Q. 568. A hole is drilled through the earth along its diameter and a stone is dropped into it. At the centre of the earth, the stone possess :

(*a*) P.E. ☐

(*b*) K.E. ☐

(*c*) Acceleration ☐

(*d*) None of the above. ☐

Q. 569. A satellite is moving round the earth in a circular orbit at a height h above the earth's surface. If the radius of the earth is R, the relation for velocity of satellite :

(*a*) $\frac{2GM}{(R+h)^2}$ ☐

(*b*) $\frac{GM}{Rh}$ ☐

(*c*) $\frac{2GM}{R-h}$ ☐

(d) $\frac{GM}{R-h}$ ☐

Q. 570. The escape velocity from earth's surface is

(a) equal to that from moon's surface ☐
(b) less than that from moon's surface ☐
(c) greater than that from moon's surface ☐
(d) equal to that from sun's surface. ☐

Q. 571. Moon has no atmosphere because :

(a) it doesn't have population ☐
(b) its surface isn't smooth ☐
(c) it is quite far off ☐
(d) root mean square velocity of all gases is more than their escape velocity from moon's surface. ☐

Q. 572. The acceleration due to gravity :

(a) has the same value everywhere in space ☐
(b) has the same value everywhere on the earth ☐
(c) varies with lattitude on the earth ☐
(d) is greater on the moon owing to its smaller diameter. ☐

Q. 573. If a planet existed whose mass was twice that of the earth and whose radius was three times greater, a 10-kg mass on its surface would weigh :

(a) 2.2 n ☐
(b) 4.4 n ☐
(c) 6.7 n ☐
(d) 13.3 n. ☐

Q. 574. A solid lead cylinder of radius R, a solid aluminium cylinder of radius *R*, a hollow lead cylinder of Radius *R*/2, and a solid lead sphere of radius *R* all start rolling down an inclined plane at the same time. The one that reaches the bottom first is the

(a) solid lead cylinder ☐
(b) solid aluminium cylinder ☐
(c) hollow lead cylinder ☐
(d) solid lead sphere. ☐

Q. 575. The total angular momentum of a system of particles

(a) remains constant under all circumstances ☐

(*b*) changes when a net external force acts upon the system ☐

(*c*) changes when a net external torque acts upon the system ☐

(*d*) none of the above. ☐

Q. 576. Which of the following instruments may be used to measure A.C. ?

(*a*) Galvanometer ☐

(*b*) Hot-wire ammeter ☐

(*c*) Moving coil galvanometer ☐

(*d*) None of these. ☐

Q. 577. Current used for electrolysis is

(*a*) A.C. ☐

(*b*) D.C. ☐

(*c*) Both of these ☐

(*d*) None of these. ☐

Q. 578. To use an electric lamp of low voltage on A.C. mains of higher voltage we use

(*a*) transistor ☐

(*b*) resistor ☐

(*c*) transformer ☐

(*d*) triode. ☐

Q. 579. Which of the following represents an ohm ?

(*a*) Volt/ampere ☐

(*b*) Joule/second ☐

(*c*) Watt/ampere ☐

(*d*) Joule/coulomb. ☐

Q. 580. The circuit in Figure below is used to produce the damped oscillations shown. How is the damping reduced ?

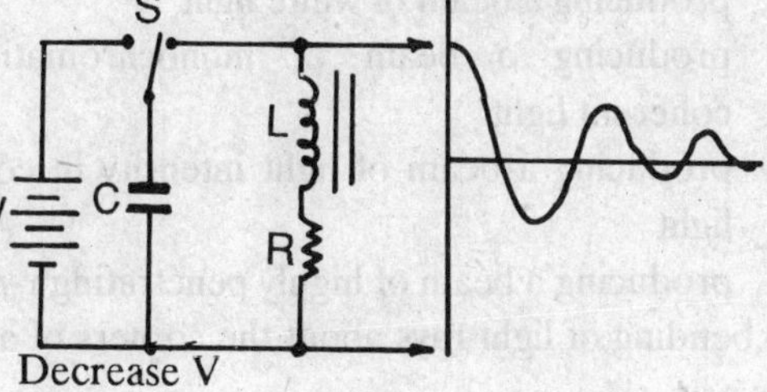

(*a*) Decrease V ☐

(*b*) Increase V
(*c*) Increase C
(*d*) Decrease R.

Q. 581. The frequency of the first line in the Lyman series for hydrogen is found to be approximately
(*a*) 2.47×10^{15} Hz
(*b*) 2.92×10^{15} Hz
(*c*) 3.08×10^{15} Hz
(*d*) 3.29×10^{15} Hz.

Q. 582. If two light nuclei are fused together in a nuclear reaction, the average energy per nucleon
(*a*) increase
(*b*) remains the same
(*c*) decrease
(*d*) cannot be determined.

Q. 583. A sample of radioactive material contains *N* radioactive nuclei at a given instant. If the half-life is 20 *s*, how many unstable nuclei remains after 1 *h* ?
(*a*) *N*/2
(*b*) *N*/4
(*c*) *N*/6
(*d*) *N*/8.

Q. 584. Which of the following nuclear reactions is possible ?
(*a*) ${}_1^1H + {}_2^3HE \rightarrow {}_2^4He$
(*b*) ${}_2^4He + {}_{13}^{27}Al \rightarrow {}_{15}^{30}P + {}_0^1n$
(*c*) ${}_1^2H + {}_{13}^{31}P \rightarrow {}_{13}^{29}Al + {}_2^4He$
(*d*) ${}_{88}^{224}Ra \rightarrow {}_{86}^{219}Rn + {}_2^4He$.

Q. 585. Laser is a device for
(*a*) producing a beam of white light
(*b*) producing a beam of monochromatic and coherent light
(*c*) producing a beam of light intensity in coherent light
(*d*) producing a beam of highly penetrating *x*-rays.

Q. 586. The bending of light rays about the corners of an obstacle is known as

(*a*) dispersion ☐
(*b*) deviation ☐
(*c*) refraction ☐
(*d*) diffraction. ☐

Q. 587. Monochromatic light from a narrow slit illuminates to narrow parallel slits producing an interference pattern on a screen. The separation between the two slits is now doubled and distance between the screen and the slits is reduced to half. The fringe width

(*a*) is doubled ☐
(*b*) becomes four times ☐
(*c*) becomes one fourth ☐
(*d*) remains the same. ☐

Q. 588. The intensity of light scattered by molecules of air in atmosphere is proportional to

(*a*) λ ☐
(*b*) λ^2 ☐
(*c*) $\frac{1}{\lambda^2}$ ☐
(*d*) $\frac{1}{\lambda^4}$. ☐

Q. 589. Two lenses of the same material can form an achromatic combination when they are

(*a*) separated by a distance equal to the product of their focal lengths. ☐
(*b*) separated by a distance equal to mean of their focal lengths. ☐
(*c*) separated by a distance equal to the sum of their focal lengths ☐
(*d*) separated by a distance equal to the difference of their focal lengths. ☐

Q. 590. Luminous flux is most closely associated with

(*a*) the source ☐
(*b*) the surface ☐
(*c*) the space between the source and the surface ☐
(*d*) none of these. ☐

Q. 591. Which of the following is *not* true when an image is formed by an object located between *C* and *F* of a concave mirror ?

(*a*) Negative magnification ☐
(*b*) Negative image distance ☐
(*c*) Inverted image ☐
(*d*) Enlarged image. ☐

Q. 592. Following figure indicates the path of white light that leaves in phase from two small sources S_1 and S_2 and travel to a point *Y* on the screen. The path difference $S_2Y - S_1Y = 11.4 \times 10^{-7}$ m select the wave length of light that gives the brightest illumination at *Y*

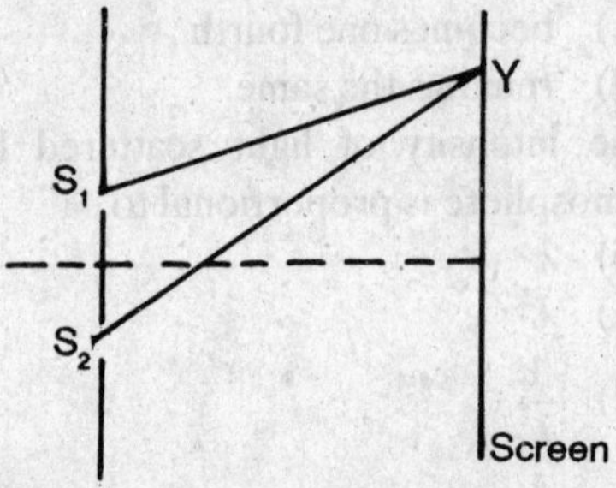

(*a*) 3.8×10^{-7} m ☐
(*b*) 5.2×10^{-7} m ☐
(*c*) 6.4×10^{-7} m ☐
(*d*) 7.6×10^{-7} m. ☐

Q. 593. A compound microscope consists of two lenses. One lens is known as objective. It is a

(*a*) convex lens of large focal length small aperture ☐
(*b*) concave lens of large focal length and large aperture ☐
(*c*) convex lens of large focal length and large aperture ☐
(*d*) convex lens of small focal length and small aperture. ☐

Q. 594. Which is the correct statement regarding the nature of light ?

(*a*) It has wave nature ☐
(*b*) It has particle nature ☐
(*c*) It has both wave and particle nature at the same time ☐

(*d*) It has wave nature at some time and particle nature at some other time. ☐

Q. 595. In a Young's double slit experiment 62 fringes are obtained in the field of view when a sodium lamp ($\lambda = 5893$ Å) is used. How many fringes will be obtained in the same field if green light ($\lambda = 5461$ Å) is used ?

(*a*) 62 ☐
(*b*) 67 ☐
(*c*) 85 ☐
(*d*) 58. ☐

Q. 596. The light waves are

(*a*) longitudinal ☐
(*b*) transverse ☐
(*c*) some times longitudinal and some times transverse ☐
(*d*) neither longitudinal nor transverse. ☐

Q. 597. Substance which are attracted by magnets and when placed in a magnetic field move from weaker to stronger parts of the field are called

(*a*) magnetic substances ☐
(*b*) paramagnetic substances ☐
(*c*) diamagnetic substances ☐
(*d*) None of these. ☐

Q. 598. Name the machine which starts working only when some initial charge is given to it.

(*a*) Electrophorous ☐
(*b*) Wimshurst machine ☐
(*c*) Glass plate machine ☐
(*d*) Van de Graaff generator. ☐

Q. 599. Substance which are attracted by the magnets and can also be magnetised are called

(*a*) paramagnetic substances ☐
(*b*) diamagnetic substances ☐
(*c*) ferromagnetic substances ☐
(*d*) None of the above. ☐

Q. 600. In a Wimshrust machine, electricity is produced by

(*a*) conduction ☐

(*b*) induction ☐
(*c*) attraction ☐
(*d*) repulsion. ☐

Section VII

Q. 601. When a ferromagnetic substance is heated its susceptibility varies inversely as the absolute temperature. This is called
(*a*) Ohm's law ☐
(*b*) Faraday's law ☐
(*c*) Curie law ☐
(*d*) None of the above. ☐

Q. 602. The temperature at which a ferromagnetism disappears and the substance becomes paramagnetic is known as
(*a*) Curie point temperature ☐
(*b*) absolute temperature ☐
(*c*) critical temperature ☐
(*d*) None of the above. ☐

Q. 603. Substance in which resultant magnetic moment is zero are called
(*a*) diamagnetic ☐
(*b*) paramagnetic ☐
(*c*) ferromagnetic ☐
(*d*) None of the above. ☐

Q. 604. Algebraic sum of currents meeting at a point is zero is a statement of
(*a*) Kirchhoff's first law ☐
(*b*) Faraday's law ☐
(*c*) Ampere's law ☐
(*d*) None of the above. ☐

Q. 605. Van de Graff's generator can start working.
(*a*) when it is given some initial charge ☐
(*b*) without giving an initial charge ☐
(*c*) with or without giving an initial charge ☐
(*d*) None of the above. ☐

Q. 606. In a closed path or mesh in a net-work of conductors carrying current the algebraic sum of the products of the resistance of and currents in the various parts of the mesh in equal to the total e.m.f. in that mesh is the statement of

(*a*) Ohm's law ☐
(*b*) Ampere's law ☐
(*c*) Kirchhoff's second law ☐
(*d*) None of the above. ☐

Q. 607. The maximum kinetic energy of photoelectrons emitted by a photosensitive metallic surface depends on

(*a*) intensity of incident radiation ☐
(*b*) frequency of incident radiation ☐
(*c*) Both of these ☐
(*d*) None of the above. ☐

Q. 608. Which of the following forces is responsible to keep nuclei together ?

(*a*) Electrostatic force only ☐
(*b*) Electrostatic and nuclear force ☐
(*c*) Nuclear force only ☐
(*d*) Electrostatic and gravitational force. ☐

Q. 609. Which of the following is used as a fuel in a nuclear reactor

(*a*) Water/heavy water ☐
(*b*) Lead/aluminium ☐
(*c*) Lead/cadmium ☐
(*d*) Plutonium/uranium. ☐

Q. 610. Photoelectric cell is a device that converts

(*a*) electrical energy into light energy ☐
(*b*) electrical energy into sound energy ☐
(*c*) light energy into electrical energy ☐
(*d*) None of the above. ☐

Q. 611. Which of the following constitute γ-rays ?

(*a*) Protons ☐
(*b*) Neutrons ☐
(*c*) Photons ☐
(*d*) None of the above. ☐

Q. 612. Which of the following constitute is β-particles ?

(*a*) Electrons
(*b*) Protons
(*c*) Photons
(*d*) None of the above.

Q. 613. In the circuit in Fig. below the ammeter reading is 0.5 A. If the resistance of the voltmeter is 1800 Ω what is the voltmeter reading ?

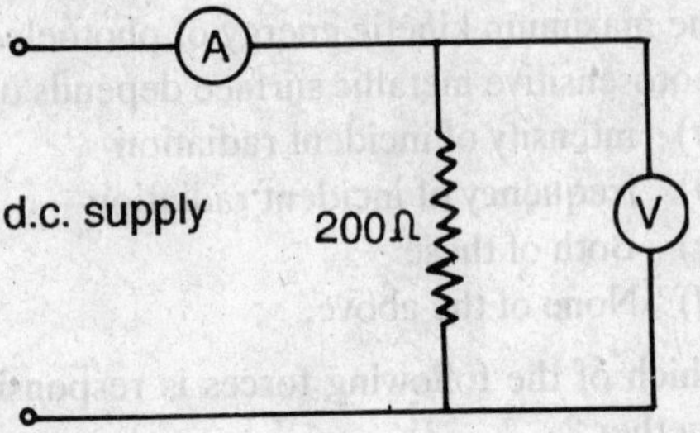

(*a*) 90 V
(*b*) 100 V
(*c*) 180 V
(*d*) 900 V.

Q. 614. Fig. below illustrates the apparatus used to calibrate a voltmeter in joules/coulomb. The ammeter reads the electric current flowing through the heating coil *XY*. The thermometers record the temperature of the water entering the leaving the glass tube.

In addition to the specific heat capacity of water and the readings on the ammeter voltmeter, and thermometers, what other measurement is required ?

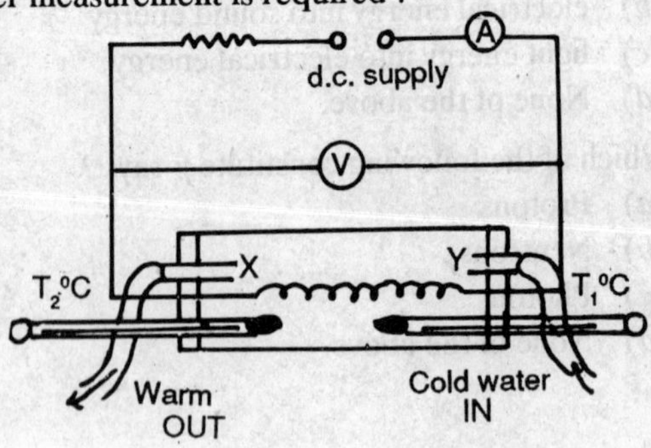

(a) Average density of water between temperatures T_1 and T_2 ☐
(b) Mass of water flowing per second through the tube ☐
(c) pressure difference of water between the ends of the tube ☐
(d) Length of the heating coil *XY*. ☐

Q. 615. Rejector circuits are
(a) Parallel resonant circuit ☐
(b) Series resonant circuit ☐
(c) Either of these ☐
(d) None of these. ☐

Q. 616. In any *LCR* circuit
(a) current lags the applied voltage ☐
(b) current leads the applied voltage ☐
(c) current some times leads and some times lags the applied voltage ☐
(d) current remains in phase with voltage. ☐

Q. 617. A lag in power factor can be controlled by
(a) inserting an inductance in parallel with circuit ☐
(b) inserting a resistance in parallel with circuit ☐
(c) inserting a condenser in parallel with circuit ☐
(d) inserting a resistane in series with circuit. ☐

Q. 618. Choose the *incorrect* statement of power factor
(a) p.f. = true watts/apparent watts ☐
(b) p.f. = apparent watts/true watts ☐
(c) p.f. = resistance/impedance ☐
(d) All the above. ☐

Q. 619. A resistance of 6 ohms is connected in series with an inductive reactance in 8 ohms and put across a 50 c/s, 200 V (r.m.s.) supply. The voltage drop along resistance is
(a) 200 V ☐
(b) 120 V ☐
(c) 160 V ☐
(d) 20 V. ☐

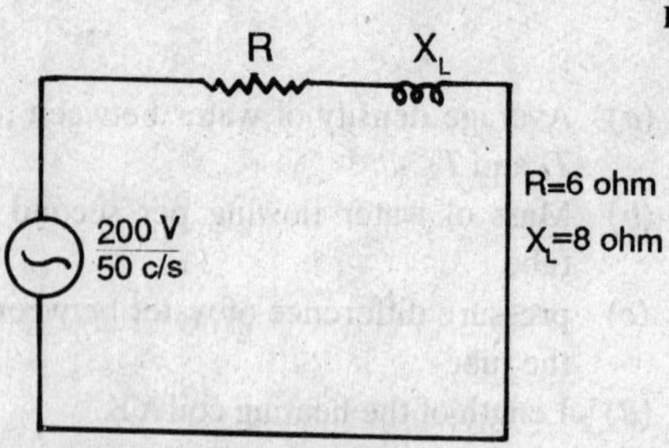

Q. 620. Deflection magnetometer is used for

(*a*) comparing the magnetic field strengths ☐
(*b*) verification of inverse square law ☐
(*c*) comparing the magnetic moments ☐
(*d*) All of these. ☐

Q. 621. Potential due to a magnetic dipole on the end side position is give by

(*a*) $\frac{1}{4\pi\mu_0}\left[\frac{M\cos\theta}{d^2}\right]$ ☐

(*b*) $\frac{1}{4\pi\mu_0}\left[\frac{M}{d^2}\right]$ ☐

(*c*) $\frac{1}{4\pi\mu_0}\left[\frac{M\cos\theta}{d^3}\right]$ ☐

(*d*) None of these. ☐

Q. 622. Which of the following is *not* correct ?

(*a*) The tangent to a line of force gives the direction of *E*-field ☐
(*b*) The tangent to a line of force also gives the magnitude of the *E*-field ☐
(*c*) The lines of force per unit cross-sectional area of field gives the magnitude of *E*-field ☐
(*d*) The *E*-field is a conservative field. ☐

Q. 623. The mariner's compass is provided with Gimbals arrangement so as to

(*a*) keep the needle always horizontal ☐
(*b*) gives a direct reading of declination ☐
(*c*) gives a direct value of dip. ☐
(*d*) None of these. ☐

Q. 624. When there is no electric field, there is no current in a circuit because

(*a*) as many electrons move in one direction as another ☐
(*b*) the electrons and ions move with the same velocity ☐
(*c*) the resistance is infinite ☐
(*d*) None of these. ☐

Q. 625. Lenz's law in electromagnetic induction follows from law of conservation of

(*a*) charge ☐
(*b*) energy ☐
(*c*) momentum ☐
(*d*) mass. ☐

Q. 626. Magnetic effect of current was discovered by

(*a*) Faraday ☐
(*b*) Fleming ☐
(*c*) Oersted ☐
(*d*) Ampere. ☐

Q. 627. In thermopile

(*a*) heat energy is converted into electrical energy ☐
(*b*) heat energy in stored ☐
(*c*) electrical energy is converted into heat energy ☐
(*d*) None of these. ☐

Q. 628. Power is transmitted from a power house on high AC voltage because

(*a*) less power is wasted at high voltage ☐
(*b*) it is very difficult to generate power at low voltage ☐
(*c*) this is a sure precaution against theft of transmission lines ☐
(*d*) the rate of transmission is faster at high voltage. ☐

Q. 629. The effective resistance offered by the resistance, capacitance and inductance in an A.C. circuit is known as

(*a*) impedence ☐
(*b*) resistance ☐
(*c*) capacitance ☐

(*d*) reactance. ☐

Q. 630. If I is the current flowing through a wire and e is the charge on electrons, then in t seconds the number of electrons are given by

(*a*) $It \,.\, e$ ☐

(*b*) $\frac{I \,.\, e}{t}$ ☐

(*c*) $\frac{I \,.\, t}{e}$ ☐

(*d*) $\frac{t}{I \,.\, e}$. ☐

Q. 631. When the temperature of metal is raised its resistance, in general,

(*a*) increase ☐

(*b*) decrease ☐

(*c*) first increase and then decreases ☐

(*d*) None of these. ☐

Q. 632. Which one gives a correct relationship between Joule and Coulomb ?

(*a*) 1 Joule = 1 $\text{Coulomb}^{-1} \times 1$ Volt ☐

(*b*) 1 Joule = 1 Coulomb $\times 1 \text{ Volt}^2$ ☐

(*c*) 1 Joule = 1 Coulomb × 1 Volt ☐

(*d*) 1 Joule = 1 $\text{Coulomb}^2 \times 1$ Volt. ☐

Q. 633. Exchange coupling is particularly important for

(*a*) diamagnetism ☐

(*b*) ferromagnetism ☐

(*c*) paramagnetism ☐

(*d*) anti-ferromagnetism. ☐

Q. 634. The phase angle between the voltage and currrent A.C. through a resistor is

(*a*) 180° ☐

(*b*) 270° ☐

(*c*) 45° ☐

(*d*) 0°. ☐

Q. 635. In purely inductive circuit, the current

(*a*) lags behind the e.m.f. by 90° ☐

(*b*) leads the e.m.f. by 90° ☐

(*c*) is in phase with e.m.f.
(*d*) may lag or lead the e.m.f.

Q. 636. Susceptance of circuit is the reciprocal of
(*a*) admittance
(*b*) resistance
(*c*) reactance
(*d*) impedence.

Q. 637. Iron loss in a dynamo arises because of
(*a*) the resistance of armature coil
(*b*) the production eddy currents
(*c*) the presence of machanical friction
(*d*) leakage of flux.

Q. 638. E.M.F. generated by a dynamo
(*a*) increases as the number of turns in the coil are increased
(*b*) decreases as the number off turns in a coil are increased
(*c*) increases as athe number of turns in a coil are decreased
(*d*) None of these.

Q. 639. Filter circuit
(*a*) eliminates D.C. component
(*b*) eliminates A.C. component
(*c*) Both the above are correct
(*d*) None of these.

Q. 640. Negative feedback
(*a*) increases stability
(*b*) decreases stability
(*c*) produces oscillation
(*d*) stops current in the tube.

Q. 641. Which type of electron wavelength is used in electron microscope ?
(*a*) Extremely short
(*b*) Short
(*c*) Large
(*d*) Moderate.

Q. 642. What is the magnification of an electron microscope ?

(*a*) 200 ☐
(*b*) 2,000 ☐
(*c*) 20,000 ☐
(*d*) 2,00,000 ☐

Q. 643. When forward bias is applied to a junction diode, it

(*a*) increases the potential barrier ☐
(*b*) decreases the potential barrier ☐
(*c*) reduces the majority carrier current to zero ☐
(*d*) reduces the minority carrier current to zero. ☐

Q. 644. When a *PN* junction is biased in the forward direction

(*a*) only holes on the *P*-type side are injected into *N*-type side ☐
(*b*) only electrons on the *N*-type side are injected ☐
(*c*) majority carriers on each side are injected ☐
(*d*) no carrier move.

Q. 645. What is Fermi energy ? It is the

(*a*) minimum energy possessed by an electron 0°K ☐
(*b*) maximum energy possessed by an electron at 0°K ☐
(*c*) maximum energy possessed by an electron at 273 K ☐
(*d*) minimum energy possessed by an electron at 273 K. ☐

Q. 646. Which of the following represents the works function of a metal ? It is the

(*a*) minimum amount of energy to be given to the slowest electron at absolute zero to allow it to escape the metal ☐
(*b*) work done in moving an electron for infinity to the metal surface ☐
(*c*) work done in moving an electron as an equipotential line in an electric field ☐
(*d*) minimum energy that must be given to the fastest moving electron at absolute zero so as to enable it to escape. ☐

Q. 647. In a triode valve, the plate resistance is 1000 ohm and the anode load resistance is 30000 ohm. If the amplification factor is 36, then voltage gain is

(*a*) 9 ☐
(*b*) 36 ☐
(*c*) 27 ☐
(*d*) 108. ☐

Q. 648. Efficiency of a half wave rectifier is

(*a*) almost negligible ☐
(*b*) more than full wave rectifier ☐
(*c*) less than full wave rectifier ☐
(*d*) None of these. ☐

Q. 649. A triode of amplification factor μ is used as an amplifier with a finite load resistor the voltage gain (A_v) is given by

(*a*) $A_v = \mu$ ☐
(*b*) $A_v < \mu$ ☐
(*c*) $A_v > \mu$ ☐
(*d*) $A_v \geq \mu$. ☐

Q. 650. There is an increase in thermionic current when the

(*a*) area of filament is decreased ☐
(*b*) area of filament is increased ☐
(*c*) temperature is decreased ☐
(*d*) work function is increased. ☐

Q. 651. The first thermionic valve which was designed by fleming in 1904 was known as

(*a*) Rectifier ☐
(*b*) Transistor ☐
(*c*) Fleming valve ☐
(*d*) None of the above. ☐

Q. 652. Arrangement of which the elements, filament or cathode and a plate kept at positive potential w.r.t. filament is called

(*a*) photo tube ☐
(*b*) diode ☐
(*c*) triode ☐
(*d*) None of the above. ☐

Q. 653. Since a diode permits the flow of current only in one direction so it can be used as

(*a*) an oscillator ☐
(*b*) a rectifier ☐
(*c*) a photodetector ☐
(*d*) None of the above. ☐

Q. 654. Transistors are made from

(*a*) plastics ☐
(*b*) metals ☐
(*c*) insulators ☐
(*d*) doped semi-conductors ☐

Q. 655. An inductance has a high resistance to AC and low DC when DC voltage sources having some AC components superimposed on it send current through an inductance to load resistance

(*a*) There is an appreciable fall in DC voltage across the load and there is only a small fall in A.C. components ☐
(*b*) There is an appreciable fall in AC voltage across the load and there is only a small fall in D.C. voltage ☐
(*c*) Both A.C. amd D.C. voltage fall by same small percentage ☐
(*d*) Both A.C. and D.C. voltage fall to nearly zero. ☐

Q. 656. Inductive time constant is given by

(*a*) L/R ☐
(*b*) R/L ☐
(*c*) L/R ☐
(*d*) $1/RL$. ☐

Q. 657. Capacity time constant is given by

(*a*) RC ☐
(*b*) R/C ☐
(*c*) C/R ☐
(*d*) $1/RC$ ☐

Q. 658. The time constant of an inductive circuit is that time during which the current in the circuit will reach with in ..... of its final value.

(*a*) 33 %
(*b*) 39 %
(*c*) 37 %
(*d*) 60 %.

Q. 659. Does the time required for the current in a particular *LR* circuit to build upto any given fraction of its equilibrium value depend upon the value of applied e.m.f. ?
(*a*) No
(*b*) Yes
(*c*) Don't know
(*d*) None of these.

Q. 660. What is the angular frequency of an *LC* circuit of $L = 10\ mh$ and $C = 10\ \mu f$
(*a*) 1×10^3 rad/sec.
(*b*) 1×10^4 rad/sec.
(*c*) 1×10^5 rad/sec.
(*d*) 1×10^6 rad/sec.

Q. 661. In a purely inductive circuit, the current
(*a*) lags behind the e.m.f. by 90°
(*b*) leads the e.m.f. by 90°
(*c*) is in phase with e.m.f.
(*d*) may lag or lead the e.m.f.

Q. 662. The r.m.s. value of e.m.f. in a circuit is given by a factor of
(*a*) 1.11
(*b*) 0.707
(*c*) 0.07
(*d*) 0.637.

Q. 663. In a series resonant circuit, the current at resonance is
(*a*) maximum
(*b*) minimum
(*c*) zero
(*d*) some times maximum some times minimum.

Q. 664. In a parallel resonant circuit, at resonance
(*a*) the current is maximum
(*b*) the voltage is maximum
(*c*) the impedance is minimum
(*d*) impedance is zero.

Q. 665. A resistance of 5 ohm is placed in series with an inductance of 200 mH and capicitance of 30 pf. If the resonant frequency is 50 c/s, what is the current at resonance, for a 100 V e.m.f. source

(*a*) 200 m amp ☐
(*b*) 20 amp ☐
(*c*) 2 amp ☐
(*d*) 200 amp. ☐

Q. 666. Which of the following relations gives the resonant frequency of a circuit ?

(*a*) $\frac{1}{2\pi\sqrt{LC}}$ ☐
(*b*) $\frac{1}{4\pi\sqrt{LC}}$ ☐
(*c*) Both the above ☐
(*d*) None of the above. ☐

Q. 667. A neon lamp requires a voltage of 60 V to 'strike'. In which circuit in Figure below is the neon lamp made to 'strike' by opening switch *S* ?

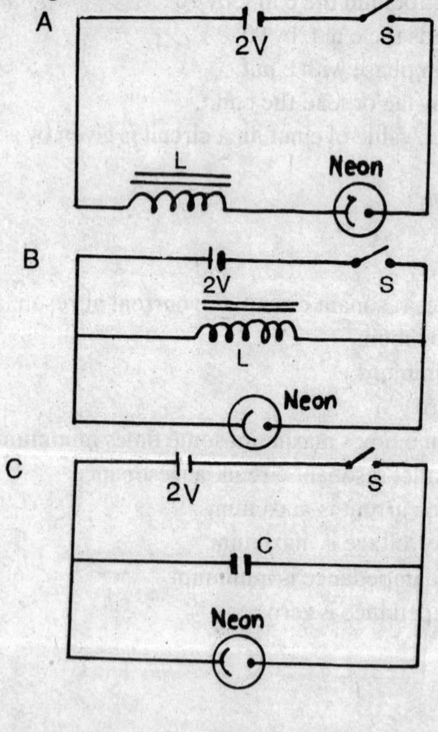

D

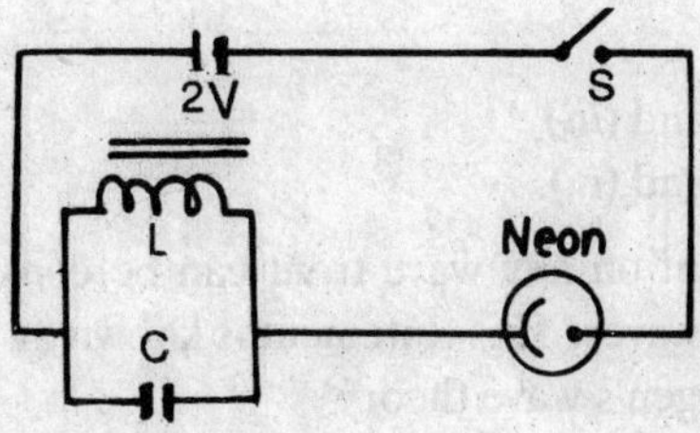

Q. 668. What is the quality for Q for a RL circuit shown below

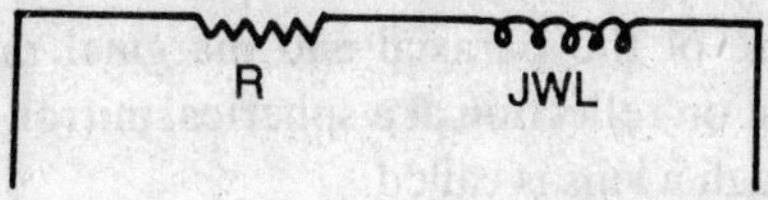

(*a*) $R/\omega L$ ☐
(*b*) $\omega L/R$ ☐
(*c*) $\omega R/L$ ☐
(*d*) ωRL. ☐

Q. 669. When the frequency of A.C. in LCR circuit is increased the impedance of LCR circuit

(*a*) increases ☐
(*b*) decreases ☐
(*c*) remains unchanged ☐
(*d*) first decreases and then increases. ☐

Q. 670. An alternating voltage of frequency ω is induced in electric circuit consisting of an inductance L and a capacitance C, connected in parallel. Then across the inductance coil

(*i*) current is maximum when $\omega^2 = \frac{1}{LC}$

(*ii*) current is minimum when $\omega^2 = \frac{1}{LC}$

(*iii*) voltage is minimum when $\omega^2 = \frac{1}{LC}$

(*iv*) voltage is maximum when $\omega^2 = \frac{1}{LC}$.

Which of the above are correct ?

(*a*) (*i*) and (*ii*) ☐
(*b*) (*i*) and (*iv*) ☐

(c) (ii) and (iii) ☐
(d) (ii) and (iv). ☐

Q. 671. Every point on any wave front can be considered as a new source of waves This statement is known as
(a) Huygen's wave theory ☐
(b) Michelson's wave theory ☐
(c) Raman's effect ☐
(d) Rayleigh's wave theory. ☐

Q. 672. The failure of the paraxial and marginal rays through the same point on reflection at a spherical mirror or refraction of light through a lens is called
(a) polarisation ☐
(b) interference ☐
(c) spherical aberration ☐
(d) diffraction. ☐

Q. 673. Which one of the following conditions a good microscopic objective should satisfy ?
(a) It should be free from chromatic aberration ☐
(b) It should be free from spherical aberration ☐
(c) A good deal of light should enter it because the object to be viewed is very small ☐
(d) All of the above. ☐

Q. 674. A *wave front* is an imaginary surface where
(a) phase changes with constant rate in all directions along the surface ☐
(b) phase changes with rate of $2\pi/\lambda$ per unit length in all directions along the surface ☐
(c) constant phase difference is always maintained ☐
(d) phase is always same for all points. ☐

Q. 675. In Young's experiment one slit is covered with the blue filter and the other slit with yellow. The interference pattern will look
(a) blue ☐
(b) yellow ☐
(c) green ☐
(d) no interference pattern will be seen. ☐

Q. 676. If white light is incident on a diffraction gratting, the light that will be deviated from the central image will be

(*a*) indigo ☐
(*b*) violet ☐
(*c*) red ☐
(*d*) crimson. ☐

Q. 677. If the whole of Young's experiment is kept in water which of the following changes will occur ?

(*a*) The fringe width will increase ☐
(*b*) The fringe width will decrease ☐
(*c*) Fringe width will remain unchanged ☐
(*d*) No fringe will be seen. ☐

Q. 678. Which is the most essential condition for producing interference ?

(*a*) The waves emitted by the two sources should either have the same phase or a constant difference in phase ☐
(*b*) The two sources must be very close to each other ☐
(*c*) The two sources be very narrow ☐
(*d*) All of these. ☐

Q. 679. Choose the correct statement, when in a Young's experiment of interference a torch is used instead of a sodium lamp

(*a*) Fringes will be produced as usual ☐
(*b*) Fringes will be produced for a moment and then disappear ☐
(*c*) Fringes wil not be produced ☐
(*d*) Fringes produced will be coloured. ☐

Q. 680. What is a "Zoom lens " ?

(*a*) It is a lens having a fixed focal length ☐
(*b*) It is a lens having variable focal length ☐
(*c*) It is a lens used in radii telescopes ☐
(*d*) None of these. ☐

Q. 681. Two sources placed at a distance '*d*' apart give interference pattern whose fringe width is found to be ω. If the sources are brought closer to a distance $d/2$, the fringe width of the interference pattern will

(*a*) undergo no change ☐
(*b*) becomes $\omega/2$ ☐

(c) becomes $3\omega/2$ ☐

(d) becomes 2ω. ☐

Q. 682. In Huygen's eye-piece

f_1 = focal length of eye lens

$3f_1$ = focal length of field lens

$2f_1$ = distance between the eye lens and field lens

The focal length F of the equivalent lens is given by the expression

(a) $\frac{2}{3}f_1$ ☐

(b) $\frac{3}{2}f_1$ ☐

(c) $\frac{\sqrt{2}}{3}f_1$ ☐

(d) $\frac{3}{\sqrt{2}}f_1$. ☐

Q. 683. The resolving power of an instrument is determined by the

(a) magnification ☐

(b) focal length of the objective lens ☐

(c) diameter of the objective lens ☐

(d) None of these. ☐

Q. 684. Which of the following properties of the wave is independent of others ?

(a) Amplitude ☐

(b) Velocity ☐

(c) Frequency ☐

(d) Wavelength ☐

Q. 685. When Newton's rings are produced by means of reflected sodium light, the diameter of dark rings is

(a) inversely proportional to the square root of natural numbers ☐

(b) directly proprtional to the square root of natural numbers ☐

(c) inversely proportional to natual numbers ☐

(d) directly proportional to natural number. ☐

Q. 686. What is the limitation to the use of thoriated tungsten emitter?

(a) Deactivation because of nagative ion bombard-ment ☐

(*b*) The temperature of operation of the tube ☐
(*c*) Deactivation because of positive ion bombardement ☐
(*d*) Can not be used at low voltages. ☐

Q. 687. Select the material that could be used for a high vacuum, high voltage tube :
(*a*) thoriated tungsten ☐
(*b*) tungsten ☐
(*c*) copper ☐
(*d*) cesium. ☐

Q. 688. Select the one that is *not* an advantage of oxide-coated cathode.
(*a*) It has a long life of several thousand years ☐
(*b*) It gives tremendous outputs at plused operations ☐
(*c*) It has high cathode efficiency ☐
(*d*) It can be used with very high voltage. ☐

Q. 689. When does a vacuum diode conduct ? It conducts when plate
(*a*) is negative with respect to cathode ☐
(*b*) is positive with respect to cathode ☐
(*c*) plate and cathode are at the same potential ☐
(*d*) resistance is less. ☐

Q. 690. A gas diode consists of
(*a*) anode only ☐
(*b*) cathode only ☐
(*c*) gas in tube only ☐
(*d*) All are true. ☐

Q. 691. On which of the following the saturation current depend in a diode valve ?
(*a*) Plate voltage ☐
(*b*) Temperature of filament ☐
(*c*) Neither of these ☐
(*d*) Both of these. ☐

Q. 692. Why are electrons tubes called valves ?
(*a*) Because they are evacuated ☐
(*b*) Because they allow flow of current in one direction ☐
(*c*) Because they are used as rectifiers ☐

(*d*) Because they are tubular in shape. ☐

Q. 693. Choose the correct statement.

(*a*) Diode acts as a rectifier ☐
(*b*) Transistor acts as an amplifier ☐
(*c*) Both (*a*) and (*b*) ☐
(*d*) None of these. ☐

Q. 694. Zener diode is used for

(*a*) rectification ☐
(*b*) amplification ☐
(*c*) stabilization ☐
(*d*) None of these. ☐

Q. 695. Crystal diodes as compared to vaccuum diodes are

(*a*) smaller in size ☐
(*b*) lighter ☐
(*c*) efficient ☐
(*d*) All of these. ☐

Q. 696. Which of the following represent a typical curve that shows a variation of plate current with grid potential for a particular plate potential

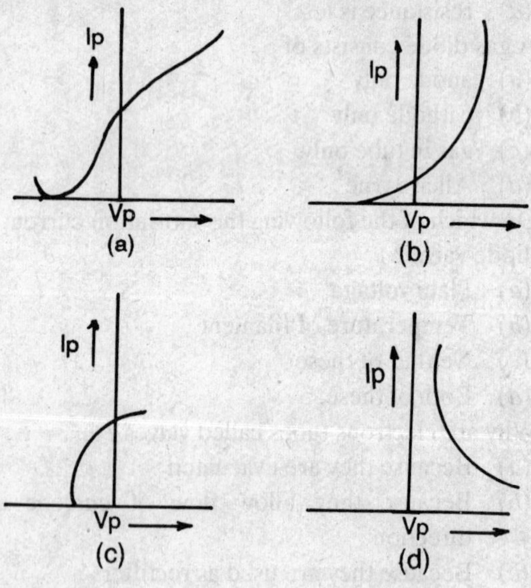

Q. 697. Where is the velocity of electrons maximum in a doide ?

(*a*) Near the cathode ☐
(*b*) Near the anode ☐
(*c*) In the space in between the two electrodes ☐
(*d*) It is same through out the tube. ☐

Q. 698. Which of the following is in accordance with Langmuir-Child Law ?

(*a*) Plate current varies as the cube of the plate potential ☐
(*b*) Plate current varies as the three-havels of the plate potential ☐
(*c*) Plate current varies as the square of the plate potential ☐
(*d*) Plate current varies as the three-quarters of plate potential. ☐

Q. 699. Which is *not* an element of triode ?

(*a*) Anode ☐
(*b*) Cathode ☐
(*c*) Control grid ☐
(*d*) Suppersser grid. ☐

Q. 700. A diode can be used for making

(*a*) an amplifier ☐
(*b*) an oscillator ☐
(*c*) a rectifier ☐
(*d*) Nothing. ☐

Section VIII

Q. 701. Plate characteristics of a triode are ploted between

(*a*) plate current and grid current ☐
(*b*) plate voltage and grid current ☐
(*c*) plate voltage and plate current ☐
(*d*) grid voltage and plate voltage. ☐

Q. 702. A hot wire ammeter uses the principle of

(*a*) defection produced in a coil of wire placed in a magnetic field when a current flows through it ☐

(*b*) expansion produced in a wire due to the heat generated in it when a current flows ☐

(*c*) change of resistance of a wire with temperature ☐

(*d*) thermoelectricity. ☐

Q. 703. A metal pendulum is attached to a clock. The clock keeps correct time at 20°C. At 40°C, the clock

(*a*) gains time ☐

(*b*) loses time ☐

(*c*) keeps correct time ☐

(*d*) None of these. ☐

Q. 704. At constant temperature, when the pressure of a given mass of gas increases by 4 times, its volume increases

(*a*) 4 times ☐

(*b*) 2 times ☐

(*c*) 1/2 times ☐

(*d*) 1/4 times. ☐

Q. 705. The atomic heat of an element is

(*a*) $\frac{\text{specific heat}}{\text{atomic heat}}$ ☐

(*b*) specific heat × atomic heat ☐

(*c*) specific heat × (atomic weight)^2 ☐

(*d*) specific heat × molecules weight. ☐

Q. 706. If c is the mean square velocity (R.M.S. velocity) of a gas and c_1, c_2, c_3, c_4 are the velocities of the individual molecules, then the relation between them is

(*a*) $c^2 = \frac{c_1^2 + c_2 + c_3^2 + \ldots + c_n^4}{n}$ ☐

(*b*) $c = \frac{c_1 + c_2 + c_3 + \ldots + c_n}{n}$ ☐

(*c*) $\sqrt{c} = \frac{\sqrt{c_1 + c_2 + c_3 + \ldots + c_n}}{n}$ ☐

(*d*) None of these. ☐

Q. 707. The temperature where Joule Thomson effect is zero is known as

(*a*) critical temperature ☐
(*b*) zero temperature ☐
(*c*) inversion temperature ☐
(*d*) None of the above. ☐

Q. 708. When a clock fitted with a seconds pendulum and taken to hill, it

(*a*) gains times ☐
(*b*) loss times ☐
(*c*) somtimes gains times and sometimes loss times ☐
(*d*) keeps correct time ☐

Q. 709. When a salt dissolved in pure water, the vapour pressure of the solution decreases. Hence in order to make the solution boil, the temperature of the solution should be increased to a value which is

(*a*) the boiling point of pure water ☐
(*b*) below the boiling point of pure water ☐
(*c*) above the boiling point of pure water ☐
(*d*) infinity. ☐

Q. 710. Thermionic emission in a tube determines the value of

(*a*) space charge limited current ☐
(*b*) temperature scattered value of the current ☐
(*c*) voltage saturated value of the current ☐
(*d*) None of these. ☐

Q. 711. Which one of the following is called mutual characteristics of triode ?

(*a*) Graph between plate current and plate voltage ☐
(*b*) Graph between plate current and grid voltage ☐
(*c*) Graph between plate voltage and grid voltage ☐
(*d*) Graph between grid current and grid voltage. ☐

Q. 712. Which of the following is *not* a characteristic of an ideal diode?

(*a*) Forward resistance zero. ☐
(*b*) Reverse resistance zero. ☐
(*c*) Both of these. ☐

(d) None of these. □

Q. 713. Which of the following gives the dynamic plate resistance for small signal operation of a diode ?

(a) $r_p = \frac{dI_p}{dv_p}$ □

(b) $r_p = dv_p . dI_p$ □

(c) $r_p = \frac{dv_p}{dI_p}$ □

(d) None of these. □

Q. 714. Which of the following gives amplification factor of a triode ?

(a) $\frac{dv_p}{dI_p}$ □

(b) $\frac{dv_p}{dv_g}$ □

(c) $-\frac{dv_p}{dv_g}$ □

(d) $\frac{dI_p}{dv_g}$. □

Q. 715. Why is an oxide coated filament used in vacuum tube ?

In vacuum tube oxide coated filament is used because it

(a) has a longer life □

(b) can withstand high voltage □

(c) emits electrons at low temperature □

(d) reduces the effect of space charge. □

Q. 716. What is the polarity of the *p*-material of a diode ?

(a) positive □

(b) Negative □

(c) Neutral □

(d) None of these. □

Q. 717. What type of charge carrier is a hole ?

(a) Negative □

(b) Positive □

(c) Neutral □

(d) None of these. □

Q. 718. In which direction do the holes move ?

(a) In the direction of flow current □

(*b*) In the direction of flow of conventional current ☐
(*c*) Opposite to that of the flow of conventional current ☐
(*d*) Can not be predicted. ☐

Q. 719. How does the conductivity of silicon or germanium change with increase of temperature ?
(*a*) Decreases ☐
(*b*) Increases ☐
(*c*) First decreases than increases ☐
(*d*) Remains unchanged. ☐

Q. 720. Select the one that is not a donor ?
(*a*) P ☐
(*b*) Sb ☐
(*c*) As ☐
(*d*) In. ☐

Q. 721. The first patented diode used as a detector for radio signals is credited to
(*a*) Edison ☐
(*b*) de Forest ☐
(*c*) Fleming ☐
(*d*) Thomson. ☐

Q. 722. In the operation of the *PNP*-transistor compared to that of a triode, the *N* base acts as the
(*a*) emitter ☐
(*b*) cathode ☐
(*c*) grid ☐
(*d*) plate. ☐

Q. 723. The picture or TV scene to be transmitted must be scanned because in TV transmission light waves are first changed into
(*a*) sound waves ☐
(*b*) electric fluctuations ☐
(*c*) magnetic fluctuations ☐
(*d*) None of these. ☐

Q. 724. In the operation of *PNP*-transistor compared to that of a triode, the *N* base acts as
(*a*) emitter ☐

(*b*) cathode ☐
(*c*) grid ☐
(*d*) plate. ☐

Q. 725. A d.c and a.c voltage are applied in turn across the terminals of an enclosed box, with these results.

(*i*) No direct current flows.
(*ii*) An alternating current flows and increases as the frequency increases.

Which of the boxes in Fig. below is used ?

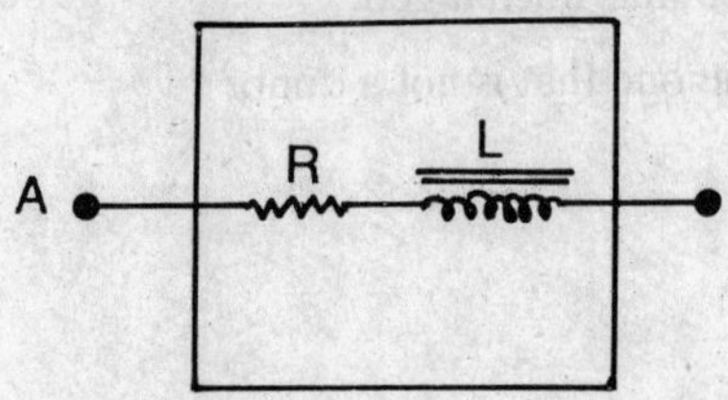

(*a*) ***A.*** ☐

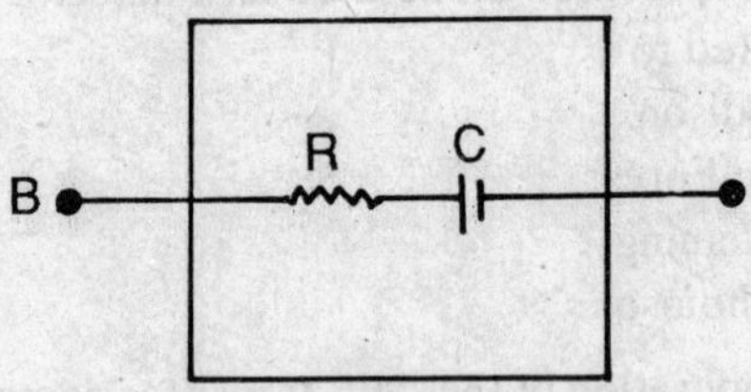

(*b*) ***B.*** ☐

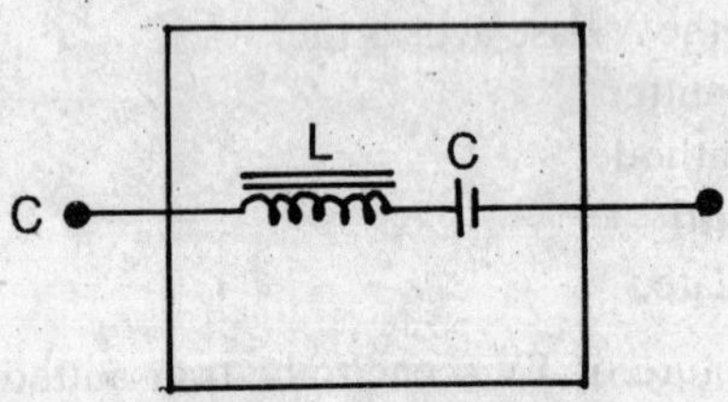

(*c*) ***C.*** ☐

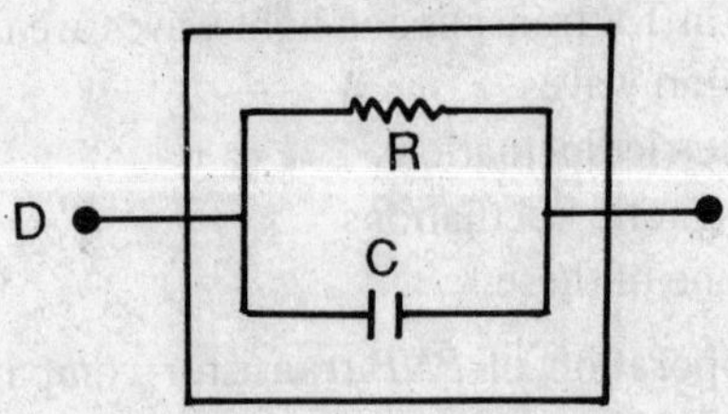

(*d*) ***D.*** ☐

Q. 726. The graph in Fig. below shows one cycle of an alternating current. If this current flows in an inductor, at which instant of the current cycle is the voltage induced in the inductor greatest ?

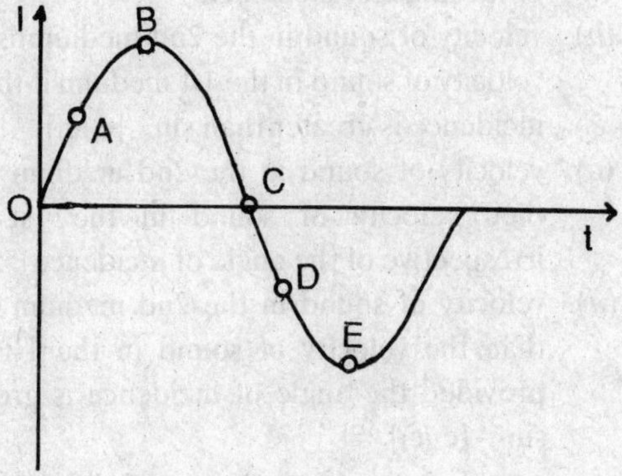

(*a*) *A* ☐
(*b*) *B* ☐
(*c*) *C* ☐
(*d*) *D*. ☐

Q. 727. In the circuit in Fig. below *V* is a high resistance voltmeter and *A* is a low resistance ammeter. Switch *S* is open.
What is effect on the voltmeter and ammeter readings when switch *S* is now closed ?

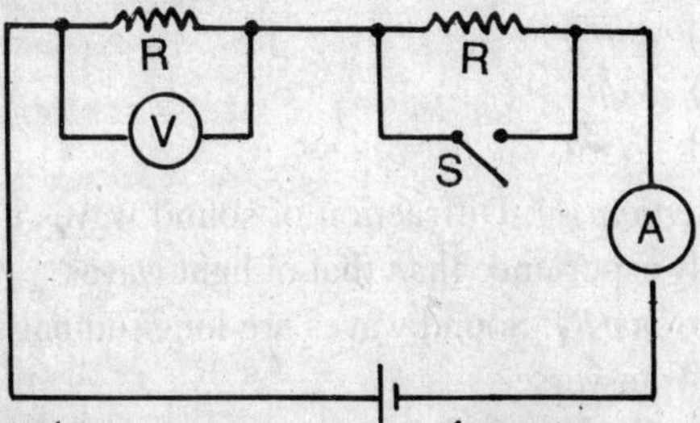

	Voltmeter reading	*Ammeter reading*	
(*a*)	Increases	Increases	☐
(*b*)	Decreases	Increases	☐
(*c*)	Increases	Stay the same	☐
(*d*)	Decreases	Decreases.	☐

Q. 728. Sound waves travelling from one medium to the other will undergo total internal reflection if

(*a*) velocity of sound in the 2nd medium is less than velocity of sound in the 1st medium irrespective of the angle of incidence ☐

(*b*) velocity of sound in the 2nd medium is less than velocity of sound in the 1st medium if the angle of incidence is greater than $\sin^{-1}[c_2/c_1]$ ☐

(*c*) velocity of sound in the 2nd medium is greater then velocity of sound in the 1st medium, irrespective of the angle of incidence ☐

(*d*) velocity of sound in the 2nd medium is greater than the velocity of sound in the 1st medium, provided the angle of incidence is greater than $\sin^{-1}[c_2/c_1]$. ☐

Q. 729. The ratio of velocity of sound through a gas to the r.m.s velocity of the gas molecules is

(*a*) γ ☐

(*b*) $\gamma/3$ ☐

(*c*) $\sqrt{\gamma}$ ☐

(*d*) $\sqrt{\gamma/3}$. ☐

Q. 730. The velocity of longitudinal waves through a solid is given by the expression

(*a*) p/q ☐

(*b*) $\sqrt{q/p}$ ☐

(*c*) $\sqrt{qp}$ ☐

(*d*) $\sqrt{p/q}$. ☐

Q. 731. *Assertion(A)*: Diffraction of sound waves is more evident in daily experience than that of light waves.

Reasons(R): Sound waves are longitudinal, while light waves are transverse.

(*a*) Both *A* and *R* are correct and *R* is the correct explanation of *A* ☐

(*b*) Both *A* and *R* are correct but *R* is not the correct explanation of *A* ☐

(*c*) *A* is correct but *R* is incorrect ☐

(*d*) *A* is incorrect but *R* is correct. ☐

Q. 732. A body moves with simple harmonic motion so that its speeds when 2 cm and 3 cm from the centre of the motion are $3cm^{-1}$ and $2cm^{-1}$ respectively. With 5/2 cm from the centre, its speed in cm^{-1} is

(*a*) 12/5 ☐
(*b*) 49/20 ☐
(*c*) 5/2 ☐
(*d*) $(27/4)^2$. ☐

Q. 733. Sound waves are classified as longitudinal because

(*a*) they require a material medium to pass ☐
(*b*) the particle displacements take place along the line of propagation ☐
(*c*) the pressure variations and the particle displacements are out of phase ☐
(*d*) they always originate from a vibrating source. ☐

Q. 734. Resonance is said to take place when

(*a*) forcing power is maximum ☐
(*b*) forcing frequency is equal to natural frequency of the forced body ☐
(*c*) forcing amplitude in maximum ☐
(*d*) the damping coefficient of the forced body is minimum. ☐

Q. 735. When two simple harmonic motions of different amplitudes are compounded on a particle at right angle with a phase difference of A, the nature of the Lissajou's figure described by the particle will be

(*a*) a straight line of negative slope ☐
(*b*) a straight line of positive slope ☐
(*c*) a curve ☐
(*d*) an ellipse. ☐

Q. 736. When a sound wave passes through medium, the maximum velocity of a particle of the medium through which the wave passes is given by

(*a*) frequency × amplitude ☐
(*b*) $\frac{\text{frequency}}{2\pi} \times$ amplitude ☐
(*c*) $2\pi \times$ frequency × amplitude ☐

(*d*) $(\text{frequency})^2 \times$ amplitude. ☐

Q. 737. The velocity of longitudnal waves through gas at a given temperature

(*a*) is unaffected by change in its pressure ☐
(*b*) increases with increase in its pressure ☐
(*c*) decreases with increase in its pressure ☐
(*d*) does not vary in any of the aforesaid ways. ☐

Q. 738. At a given temperature, sound takes 2 seconds to travel from a point *A* to a point *B* through air. When the Kelvin temprature of the medium is doubled, the same sound takes for the same travel the time of

(*a*) 1 second ☐
(*b*) $\sqrt{2}$ seconds ☐
(*c*) $2\sqrt{2}$ seconds ☐
(*d*) 4 seconds. ☐

Q. 739. A fathameter is an instrument used to measure

(*a*) velocity of sound ☐
(*b*) frequency of sound ☐
(*c*) depth of sea ☐
(*d*) wavelength of sound. ☐

Q. 740. If the earth stopped rotating, the weight of objects at the equator would

(*a*) be greater ☐
(*b*) be same as before ☐
(*c*) be less ☐
(*d*) vary with latitude. ☐

Q. 741. The man in exercise 3 could have reached the same destination if he had headed in. What is single direction from the start ?

(*a*) 45° east of north ☐
(*b*) 22° east of north ☐
(*c*) 50° east of north ☐
(*d*) 63° east of north. ☐

Q. 742. A satellite travels in a circular orbit at a speed of 20,000 km/h to stay at a constant attitude. To escape from the earth, the speed would have to be increased to

(a) 28,000 km/h
(b) 21,000 km/h
(c) 40,000 km/h
(d) 64,000 km/h.

Q. 743. The gravitational force with which the earth attracts the moon
(a) is less than the force with which the moon attracts the earth
(b) is the same as the force with which the moon attracts the earth
(c) is more than the force the moon attracts earth
(d) varies with the phase of the moon.

Q. 744. If the earth stopped rotating, the weigths of objects at either pole would
(a) be less
(b) be greater
(c) be the same as before
(d) vary the latitude.

Q. 745. The gravitational force between two bodies does not depend upon
(a) the product of their masses
(b) their separation
(c) the sum of their masses
(d) the constant of gravitation.

Q. 746. A satellite is orbiting close to the earth. In order to make it move to infinity, its orbital speed is to be increased by
(a) 20 %
(b) 10 %
(c) 41.4 %
(d) 100 % .

Q. 747. When a ball on a string moves in a vertical circle, the tension on the string is greatest when
(a) string is made longer
(b) speed of the ball increased
(c) the ball is at the highest point
(d) speed of the ball decreased

Q. 748. When a car is travelling at constant speed around a circular track, a quantity that is constant but not zero is its

(*a*) acceleration ☐
(*b*) angular velocity ☐
(*c*) velocity ☐
(*d*) angular acceleration. ☐

Q. 749. When the angular position of the swing is 45°, its linear velocity is

(*a*) negative and increasing ☐
(*b*) positive and decreasing ☐
(*c*) positive and increasing ☐
(*d*) negative and decreasing. ☐

Q. 750. Due to redistribution of a mass of rotating body if its moment of inertia increases then its angular velocity

(*a*) decreases ☐
(*b*) remains unchanged ☐
(*c*) increases ☐
(*d*) None of the above. ☐

Q. 751. If I and E are the moment of inertia and rotational kinetic energy of body respectively then its angular momentum L is calculated form

(*a*) $\frac{2E^2}{2I}$ ☐
(*b*) $2E^2I$ ☐
(*c*) $\sqrt{2EI}$ ☐
(*d*) $2EI$. ☐

Q. 752. Physical quantity not directly involved in rotatory motion is

(*a*) moment of inertia ☐
(*b*) mass ☐
(*c*) angular velocity ☐
(*d*) torque. ☐

Q. 753. Vibration of a body is given by the differential equation

$$\frac{d^2x}{dt^2} + \omega^2 x = 0$$

The natural frequency is given by

(*a*) ω ☐

(b) $2\pi/\omega$ ☐
(c) $\omega/2\pi$ ☐
(d) ω^2. ☐

Q. 754. A shell is fired from a gun, whose barrel is inclined at an angle to the horizon. During the flight, the shell explodes in air into fragments. The centre of mass of different fragments will follow

(a) a horizontal straigth line ☐
(b) a vertical straight line ☐
(c) a parabola ☐
(d) an unpredictable path. ☐

Q. 755. Two bodies with velocities u_1 and u_2 collide and thereafter move with a common velocity. The coefficient of restitution between the bodies is

(a) 1 ☐
(b) zero ☐
(c) $\mu_1 + \mu_2$ ☐
(d) $\mu_1 - \mu_2$. ☐

Q. 756. The gravitational potential at a point which is at height equal to the radius of the earth (R) from the surface of the earth is

(a) $-2gR$ ☐
(b) $+2gR$ ☐
(c) $gR/2$ ☐
(d) $+gR/2$. ☐

Q. 757. A particle executing simple harmonic oscillations of frequency 100 Hz has an amplitute 0.1 cm. The velocity of amplitude of particle is

(a) 20 π cm/sec ☐
(b) 10 π cm/sec ☐
(c) 20 cm/sec ☐
(d) 10 cm/sec ☐

Q. 758. When 2 simple harmonic motions having the same amplitude A, frequency n and phase d are superimposed on a particle along the same straight line, the particle execute simple harmonic motion of

(a) amplitude A and frequency n ☐

(*b*) amplitude A and frequency $2n$ ☐
(*c*) amplitude $2A$ and frequency n ☐
(*d*) amplitude $2A$ and frequency $2n$. ☐

Q. 759. A men with a pair of dumbells in his outstretched hand stands in a revolving platform. If he drops the dumbells at any instant, thereafter the rate of revolution of platform

(*a*) decreases ☐
(*b*) increases ☐
(*c*) remains the same ☐
(*d*) None of these. ☐

Q. 760. When 2 simple harmonic motions are compounded on a particle at right angle to each other, the particle describes a circle. This implies that the 2 simple harmonic motions have

(*a*) same amplitude and phase ☐
(*b*) same amplitude but a phase difference of $\pi/2$ ☐
(*c*) same amplitude but a phase difference of π ☐
(*d*) same phase but different amplitude. ☐

Q. 761. Who discovered the phenomenon of thermoelectricity ?

(*a*) Peltier ☐
(*b*) Joule ☐
(*c*) Seeback ☐
(*d*) Thomson. ☐

Q. 762. In a thermopile

(*a*) heat energy is converted into electrical energy ☐
(*b*) heat energy is stored ☐
(*c*) electrical energy is changed into heat energy ☐
(*d*) None of these. ☐

Q. 763. In the thermocouple of pure copper and pure iron the temperature of cold junction is 0°C and the hot junction is heated from 20° to 500°C, what will be the electromotive force ?

(*a*) Increase to maximum before becoming constant ☐
(*b*) Increases with increase of temprature ☐
(*c*) First increases to maximum and then decreases to minimum ☐
(*d*) Rise to a maximum, decline to zero and then reverse direction. ☐

Q. 764. Heat is developed fastest in a wire
- (*a*) of small thickness ☐
- (*b*) of large thickness ☐
- (*c*) does not depend on thickness of wire ☐
- (*d*) None of these. ☐

Q. 765. Two resistances '*x*' and '*y*' of 15 and 25 ohms respectively are connected across a battery of V, first in parallel and then in series. The heat produced is
- (*a*) more in first case ☐
- (*b*) more in second case ☐
- (*c*) equal in both cases ☐
- (*d*) None is correct. ☐

Q. 766. The amount of heat developed in a resistor is directly proportional to
- (*a*) the square of the current only ☐
- (*b*) the resistance of the conductor only ☐
- (*c*) the time of current passage only ☐
- (*d*) The square of the current, resistance and the time of current flow. ☐

Q. 767. Neutral temperature for a thermocouple is defined as the temperature at which
- (*a*) the thermo e.m.f. changes sign ☐
- (*b*) the thermo e.m.f. is maximum ☐
- (*c*) the thermo e.m.f. is minimum ☐
- (*d*) the thermo current is maximum. ☐

Q. 768. In Sb-Bi couple, the direction of flow of current is from
- (*a*) Bi to Sb ☐
- (*b*) Sb to Bi ☐
- (*c*) Any of these ☐
- (*d*) None of these. ☐

Q. 769. If a strong current is passed across the junction from Bi to Sb a cooling is observed. This is because of
- (*a*) Seeback effect ☐
- (*b*) Thomson's effect ☐
- (*c*) Peltier's effect ☐
- (*d*) Zeeman effect. ☐

Q. 770. In elliptically polarised light
(*a*) the amplitude of vibrations changes in direction only ☐
(*b*) the amplitude of vibrations changes in magnitude only ☐
(*c*) the amplitude of vibration changes in direction as well as in magnitude ☐
(*d*) None of these. ☐

Q. 771. When your view through a microscope a thin film lying between a plane glass plate of a convex lens and irradiated with parallel beam of monochromatic light, you will find
(*a*) complete darkness ☐
(*b*) uniform brightness ☐
(*c*) field crossed by dark and bright fringes ☐
(*d*) field crossed by concentric dark and bright fringes. ☐

Q. 772. A soap bubble when viewed in white light shows colour due to
(*a*) interference ☐
(*b*) scattering ☐
(*c*) diffraction ☐
(*d*) dispersion. ☐

Q. 773. What interference will you draw if you find the intensity ratio between bright and dark fringes as 9 in a Young's double slit experiment ?
(*a*) Amplitude ratio is 3 ☐
(*b*) Amplitude ratio is 2 ☐
(*c*) Instensity at the screen due to the two slits are 5 units and 4 units respectively ☐
(*d*) All the above are ture. ☐

Q. 774. For the internal study of the structure of crystal, we make use of
(*a*) reflection ☐
(*b*) diffraction ☐
(*c*) polarisation ☐
(*d*) refraction. ☐

Q. 775. In an interference pattern minima has zero intensity when the disturbances superimposed have

(*a*) unequal amplitudes ☐
(*b*) equal amplitudes ☐
(*c*) unequal phases ☐
(*d*) None of these. ☐

Q. 776. When two waves superimpose, then in the region of superimposition

(*a*) energy is conserved ☐
(*b*) mass is conserved ☐
(*c*) intensity of light is same every where ☐
(*d*) All the above are ture. ☐

Q. 777. A fringe is a path is

(*a*) constant amplitude ☐
(*b*) constant phase ☐
(*c*) same wave length ☐
(*d*) None of these. ☐

Q. 778. The interference pattern obtained by making use of two candles is

(*a*) a sustained interference pattern ☐
(*b*) temporary interference pattern ☐
(*c*) no interference pattern ☐
(*d*) All are true. ☐

Q. 779. A Bi-prism consists of

(*a*) two parallel glass plates ☐
(*b*) two acute angled prisms ☐
(*c*) two obtuse angled prisms ☐
(*d*) None of these. ☐

Q. 780. The interference can occur only if the two sources of light

(*a*) have same frequency ☐
(*b*) have any frequency ☐
(*c*) have frequencies very near to each other ☐
(*d*) None of these. ☐

Q. 781. In Fresnel's biprism interference is produced by

(*a*) division of amplitude method ☐
(*b*) division of wave front method ☐

(c) addition of amplitude method ☐
(d) None of these. ☐

Q. 782. When interference of light occurs
(a) energy is created in the region of maximum intensity ☐
(b) destroyed in the region of minimum intensity ☐
(c) energy is neither created nor destroyed, it is only redistributed ☐
(d) All are true. ☐

Q. 783. A chromatic fringe can be obtained with
(a) white light ☐
(b) invisible light ☐
(c) in coherent light ☐
(d) None of these. ☐

Q. 784. Which of the following is *correct* for coherent sources ?
(a) The waves have equal energy ☐
(b) The waves originating from these have costant phase difference ☐
(c) The waves originating from these have same frequency ☐
(d) All the above. ☐

Q. 785. Newton's rings are produced by
(a) division of wave front method ☐
(b) division of amplitude method ☐
(c) addition of amplitude method ☐
(d) None of these. ☐

Q. 786. The main advantage of a grating over Young's apparatus is the
(a) sharpness of the bright lines ☐
(b) absence of dark fringes ☐
(c) absence of bright fringes ☐
(d) greater deviation of light. ☐

Q. 787. In a diffraction pattern, the width of any fringe is
(a) directly proportional to slit width ☐
(b) inversely proportional to slit width ☐
(c) independent of slit width ☐

(*d*) None of these. ☐

Q 788. The thermal inertia of a thermodynamic system is know is

(*a*) its enthalpy ☐
(*b*) its entropy ☐
(*c*) its isothermal condition ☐
(*d*) its adiabatic condition. ☐

Q. 789. The phenomenon of land and sea breeze is as a resut of

(*a*) conduction ☐
(*b*) covection ☐
(*c*) radiation ☐
(*d*) None of these. ☐

Q. 790. If a gas in a closed container is allowed to expand without supplying any energy, whcih of the following statements is about the behaviour of molecules will be *time* ?

I. They will collide less frequency with each other
II. They will more faster
III. They will move more slowly
IV. The collisions will become inelastic instead of elastic
V. They won't collide so frequently with the walls of the container

(*a*) Only I ☐
(*b*) I, II, IV, V ☐
(*c*) I, II, III, V ☐
(*d*) I, III, V. ☐

Q. 791. According to the law of equipartition of energy the total energy of any dynamical system in thermal equilibrium is equally divided amongst all the degrees of freedom available. The energy associated with each molecule per degree of freedom is (k is Boltzmann constant, T is absolute temperature)

(*a*) $3/2\,kT$ ☐
(*b*) $1/2\,kT$ ☐
(*c*) kT ☐
(*d*) None of these. ☐

Q. 792. Gases exert pressure on the walls of the containing vessel because as the gas molecules

(*a*) possess momentum ☐
(*b*) have finite volume ☐
(*c*) collide with each other ☐
(*d*) obey gas laws. ☐

Q. 793. Boyle's law is an example of a (*n*)

(*a*) Adiabatic proces ☐
(*b*) Isothermal process ☐
(*c*) Latent heat process ☐
(*d*) Mechanical process. ☐

Q. 794. The kinetic molecular theory of gases predicts that at a given temperature

(*a*) light gas molecules have lower average energies than heavy gas molecules ☐
(*b*) all of the molecules in a gas have the same average energy ☐
(*c*) all of the molecules in a gas have the same average speed ☐
(*d*) light gas molecules have higher average energies than heavy gas molecules. ☐

Q. 795. Absolute zero may be regarded as that temperature at which

(*a*) all gases become liquids ☐
(*b*) water freezes ☐
(*c*) all substances are solid ☐
(*d*) molecular motion in a gas would cease. ☐

Q. 796. The pressure of a gas is directly proportional to

(*a*) mean velocity fo the molecules ☐
(*b*) velocities of individual molecules ☐
(*c*) root mean square velocity of the molecules ☐
(*d*) mean square velocity of the molecules. ☐

Q. 797. For a perfect gas, whose molecule have n degree of freedom, the ratio of specific heat at constant pressure and specific heat at constant volume

(*a*) $1 - n$ ☐
(*b*) $1 + n$ ☐

(c) 1 + 2/n

(d) n.

Q. 798. At a given temperature, the average velocity of bromine gas molecules is

(a) a little less than

(b) equal

(c) a little greater than

(d) much less than the average velocity of hydrogen gas molecules.

Q. 799. The expression for root mean square velocity is

(a) $\frac{P}{3\rho}$

(b) $\frac{P}{\rho}$

(c) $\frac{3P}{\rho}$

(d) $\frac{3P}{1/\rho}$.

Q. 800. The kinetic molecular theory of gases is based upon

(a) assumptions that are considered dubious

(b) experimental observation of molecules

(c) assumptions that are considered valid because the results of the theory agree with experimental data

(d) assumptions that are considered valide because they are in accord with our intuitive feelings.

Section IX

Q. 801. Interferometers are based on the principle of

(a) diffraction

(b) superposition

(c) interference

(d) None of these.

Q. 802. The phenomenon of interference explained by

(*a*) Raman's spectra ☐
(*b*) Newton's rings ☐
(*c*) Complex effect ☐
(*d*) Emission spectra. ☐

Q. 803. Light is polarised to the maximum when light is incident on a glass surface at an angle of

(*a*) 57° ☐
(*b*) 47° ☐
(*c*) 67° ☐
(*d*) All are true. ☐

Q. 804. Which of the following terms is used for light which has same property in all directions ?

(*a*) Polarised light ☐
(*b*) Unpolarised light ☐
(*c*) Monochromatic light ☐
(*d*) None of these. ☐

Q. 805. Michelson's interferometer is based on the principle of

(*a*) division of amplitude ☐
(*b*) division of wave front ☐
(*c*) addition of amplitude ☐
(*d*) None of these. ☐

Q. 806. If the intensities of two interfering beams in Young's double slit experiment be I_1 and I_2 then the contrast between the maximum and minimum intensity is good when

(*a*) I_1 is much greater than I_2 ☐
(*b*) I_1 is much smaller than I_2 ☐
(*c*) either $I_1 = 0$ or $I_2 = O$ ☐
(*d*) $I_1 = I_2$. ☐

Q. 807. Michelson's interferrometer can be used to measure

(*a*) wave length of light ☐
(*b*) intensity of light ☐
(*c*) amplitude of disturbance ☐
(*d*) None of these. ☐

Q. 808. The plane of polarisation is the plane

(*a*) in which vibrations take place ☐

(*b*) in which no vibrations take place ☐
(*c*) in which longitudinal vibrations take place ☐
(*d*) None of the above is correct. ☐

Q. 809. Spherical aberration in a spherical mirror can be removed
(*a*) by limiting the number of rays by using a stop ☐
(*b*) by using an ellipsoidal mirror ☐
(*c*) by using a parabolic mirror ☐
(*d*) by any one of the above. ☐

Q. 810. An achromatic combination is to be obtained using a convex lens. The two lenses chosen should have
(*a*) their powers equal ☐
(*b*) their refractive indices equal ☐
(*c*) their dispersive powers equal ☐
(*d*) the product of their powers and dispersive powers equal. ☐

Q. 811. An optical device made from calcite crystal and used in many optical instrument to produce and analyse plane polarised light is known as
(*a*) poloroid ☐
(*b*) polariser ☐
(*c*) nicol prism ☐
(*d*) None of these. ☐

Q. 812. The power of lens which is the reciprocal of focal length, is also given by the formula in terms of refractive index and radius of curvature
(*a*) $(\mu-1)\left(\frac{1}{R_1}+\frac{1}{R_2}\right)$ ☐
(*b*) $\mu\left(\frac{1}{R_1}-\frac{1}{R_2}\right)$ ☐
(*c*) $\frac{\mu\,(R_1+R_2)}{R_1}$ ☐
(*d*) $(\mu-1)\left(\frac{1}{R_1}-\frac{1}{R_2}\right)$ ☐

Q. 813. A crystal plate that can produce a phase differenc of $\frac{\pi}{2}$ and path difference of $\frac{\lambda}{4}$ between ordinary ray and extra ordinary ray is known as

(a) nicol prism ☐
(b) half wave plate ☐
(c) quarter wave plate ☐
(d) None of these. ☐

Q. 814. Lens-maker's formula
(a) is applicable to very thin lens of very small aperture ☐
(b) is applicable to all forms of lens ☐
(c) apart from being applicable to very thin lenses is applicable to other lenses under certain condition ☐
(d) None of these. ☐

Q. 815. A crystal plate which can produce a phase difference of π or path difference of $\lambda/2$ between ordinary ray and extra-ordinary ray is known as
(a) Nicol prism ☐
(b) half wave plate ☐
(c) quarter wave plate ☐
(d) None of the above. ☐

Q. 816. Longitudinal chromatic aberration is
(a) dispersive power/mean focal length ☐
(b) mean focal length ☐
(c) dispersive power ☐
(d) dispersive power × mean focal length. ☐

Q. 817. In case of a lens, a pair of conjugate points free from spherical aberration and coma, is known as
(a) aplautic points ☐
(b) cardinal points ☐
(c) image points ☐
(d) All the above. ☐

Q. 818. Which of the following gives the difference between coma and spherical aberration ?
(a) Coma is measured laterally, while spherical aberration is measured longitudinally ☐
(b) The circles formed in coma are cometic, while those in spherical aberration is measured longitudinally ☐

(*c*) Coma is measured longitudinally, while spherical aberration is measured laterally ☐
(*d*) Both (*a*) and (*b*). ☐

Q. 819. The theory that light is emitted in discrete amounts of energy rather than in continuous fashion is known as

(*a*) the photoelectric effect ☐
(*b*) the quantum theory ☐
(*c*) Huygen's principle ☐
(*d*) the electromagnetic theory. ☐

Q. 820. Which of the following best applies for yellow light ?

(*a*) 640 nm ☐
(*b*) 0.4 nm ☐
(*c*) 5×10^{14} Hz ☐
(*d*) 510 A. ☐

Q. 821. If a current is flowing through a flat strip metal conductor and it is placed in a B-field going into the conductors as shown below. Which of the points '*X*' or '*Y*' will attain a higher potential ?

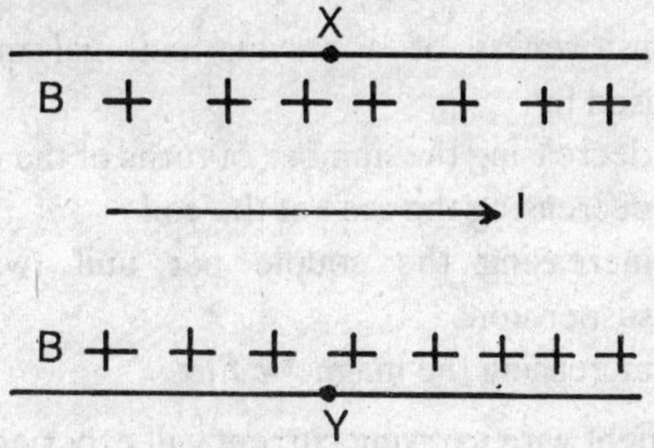

(*a*) *X* ☐
(*b*) *Y* ☐
(*c*) *X* and *Y* have same potentials ☐
(*d*) None of these. ☐

Q. 822. In a electromagnetic wave

(*a*) power is transferred along the magnetic field ☐
(*b*) power is transferred along the electric field ☐
(*c*) power is equally transferred along the electric and magnetic fields ☐

(*d*) power is transferred in direction perpendicular to both the fields. ☐

Q. 823. Which of the following can be used to accelerate neutrons ?

(*a*) Cyclotron ☐

(*b*) Synchro-cyclotron ☐

(*c*) Bevatron ☐

(*d*) High speed bombardment. ☐

Q. 824. On which of the following the time of passage of a deutron through a dee in a cyclotron depends ?

(*a*) The speed of electron ☐

(*b*) The magnetic field ☐

(*c*) The mass of deutron ☐

(*d*) None of these. ☐

Q. 825. In an induction coil, the secondary e.m.f. is

(*a*) zero at the break of the circuit ☐

(*b*) zero during make of the circuit ☐

(*c*) very high during make of the circuit ☐

(*d*) very high during break of the circuit. ☐

Q. 826. th sensitiveness of a moving coil galvanometer can be increased by

(*a*) decreasing the number of turns of the coil ☐

(*b*) decreasing the area of the coil ☐

(*c*) increasing the couple per unit twist of the suspension ☐

(*d*) increasing the magnetic flux. ☐

Q. 827. A straight wire carrying current will experience a force when placed in uniform magnetic field if

(*a*) the current and field are parallel ☐

(*b*) the current and field are at an angle ☐

(*c*) the current and field are parallel in opposite directions ☐

(*d*) None of these. ☐

Q. 828. When different parts of a metal are maintained at different temperatures and e.m.f. is observed. This can be explained by

(*a*) Stark effect ☐

(*b*) Zeeman effect ☐

(c) Peltier effect ☐
(d) Thomson's effect. ☐

Q. 829. When a direct current is passed through a junction formed of two dissimilar metals, the junction becomes warmer or cooler depending on the
(a) current direction ☐
(b) thermocouple used ☐
(c) temperature gradient ☐
(d) Piezoelectric effect. ☐

Q. 830. Thermopile is a combination of
(a) thermocouples ☐
(b) voltmeters ☐
(c) resistances in series ☐
(d) calorimeters. ☐

Q. 831. Two conducting loops of different sizes face each other a distance 'd' apart. An observer sights along their common axis. If a clockwise current *I* is suddenly established in the larger loop, what is the direction of current in the smaller loop?
(a) Clockwise ☐
(b) Anit-clockwise ☐
(c) No current will flow ☐
(d) None of these. ☐

Q. 832. According to Faraday's law, the total charge induced in a conductor that is moved in magnetic field depends upon
(a) initial magnetic flux ☐
(b) final magnetic flux ☐
(c) rate of change of magnetic flux ☐
(d) change in magnetic flux. ☐

Q. 833. Self inductance of a solenoid is
(a) directly proportional to the current flowing through the coil ☐
(b) directly proportional to its length ☐
(c) inversely proportional to area of cross-section ☐
(d) directly proportional to number of turns. ☐

Q. 834. A is singing a note and at the same time B is singing a note with exactly one eighth the frequency of the note of A. The energies of two sounds are equal, the amplitude of the note B is

(*a*) same as that of A ☐
(*b*) twice as that of A ☐
(*c*) four times as that of A ☐
(*d*) eight times as that of A. ☐

Q. 835. To produce mechanical waves, we need a source producing a disturbance and

(*a*) vacuum ☐
(*b*) electromagnetic field ☐
(*c*) elastic medium ☐
(*d*) inelastic medium. ☐

Q. 836. The intensity level in decibles of a sound whose intensity is $2x^{10-6}$ $\mu w/km^2$ is approximately

(*a*) 27 *d*B ☐
(*b*) 103 *d*B ☐
(*c*) 43 *d*B ☐
(*d*) 50 *d*B. ☐

Q. 837. The optimum reverberation time in an ordinary room is

(*a*) 1 to 2 sec ☐
(*b*) 5 to 6 sec ☐
(*c*) 10 to 15 sec ☐
(*d*) 12 to 18 sec. ☐

Q. 838. Frequencies less than 20 cycles per second are known as

(*a*) infrasonics ☐
(*b*) ultrasonics ☐
(*c*) supersonics ☐
(*d*) None of the above. ☐

Q. 839. When the mass of a string is increased to four times its original value, the velocity of the waves is

(*a*) doubled ☐
(*b*) reduced to one half ☐
(*c*) reduced to one fourth ☐
(*d*) increased to four times in original value. ☐

Q. 840. Difference in density of two media when waves are passing from one into another media always results in change in

(*a*) wave speed ☐
(*b*) wave direction ☐
(*c*) both wave speed and direction ☐
(*d*) wave frequency. ☐

Q. 841. To produce beats it is necessary to use two waves

(*a*) travelling in opposite directions ☐
(*b*) of slighty different frequencies ☐
(*c*) of equal wave length ☐
(*d*) of equal amplitudes. ☐

Q. 842. An automobile sounding its horn is moving away from an observer. The pitch of the horn's sound relative to its normal pitch is

(*a*) higher ☐
(*b*) lower ☐
(*c*) the same ☐
(*d*) higher or lower depending upon the exact frequency. ☐

Q. 843. In a longitudinal wave, the individul particles of the medium move

(*a*) in circles ☐
(*b*) in elipses ☐
(*c*) parallel to wave propagation ☐
(*d*) perpendicular to wave propagation. ☐

Q. 844. Echoes arise from

(*a*) reflection ☐
(*b*) refraction ☐
(*c*) diffraction ☐
(*d*) dispersion of sound waves. ☐

Q. 845. The Doppler effect takes place when

(*a*) a sound source is stedy only ☐
(*b*) a sound source is moving only ☐
(*c*) both the sound source and hearing person are moving ☐

(*d*) a sound source is steady and a hearing person is moving. □

Q. 846. Increased loudness produced when two bodies vibrate sympathetically is called

(*a*) an echo □
(*b*) beats □
(*c*) destructive interference □
(*d*) resonance. □

Q. 847. A water wave is an example of

(*a*) diffracted wave □
(*b*) electromagnetic wave □
(*c*) longitudinal wave □
(*d*) transverse wave. □

Q. 848. At the mean position of vibration, the velocity of the vibrating particle is

(*a*) zero □
(*b*) infinity □
(*c*) maximum □
(*d*) None of the above. □

Q. 849. The pitch of a sound is determined by its

(*a*) speed □
(*b*) frequency □
(*c*) direction □
(*d*) number of beats. □

Q. 850. A point on the crest of a transverse wave shall reach the trough after time

(*a*) $T/4$ □
(*b*) $T/2$ □
(*c*) $3T/4$ □
(*d*) T. □

Q. 851. A dise with equally-spaced holes round the edge is rotated at a constant speed. A jet of air is directed on to the hole and a musical note is heard. The pitch of this note can be lowered provided

(*a*) the pressure of air jet is decreased □
(*b*) the holes are made smaller □

(c) jet of a denser gas than air is used ☐
(d) the speed of rotation is decreased. ☐

Q. 852. Kepler modified the Copernican system by showing that the planetary orbits are
(a) elipses ☐
(b) circles ☐
(c) combinations of circles forming looped orbits ☐
(d) the same distance apart from one another. ☐

Q. 853. If there were no atmosphere, the duration of the day on the earth will:
(a) decrease ☐
(b) remain the same ☐
(c) increase ☐
(d) depend upon the weather. ☐

Q. 854. If the earth were three times farther from the sun than it is now, the gravitational force exerted on it by the sun would bbe
(a) nine times as large as it is now ☐
(b) three times as large as it is now ☐
(c) one third as large as it is now ☐
(d) one nith as large as it now. ☐

Q. 855. An artificial satellite is moving in a circular orbit about the earth. If R is the radius of the earth and h the height of the satellite above the surface of the earth, then which of the following formulae is used for the orbital velocity of the satellite ?

(a) $v = R\left[\dfrac{g}{(R-h)}\right]^{1/2}$ ☐

(b) $v = R\left[\dfrac{g}{(R-h)}\right]$ ☐

(c) $v = R\left[\dfrac{g}{(R+h)}\right]^{2}$ ☐

(d) $v = R\left[\dfrac{g}{(R-h)^{1/2}}\right].$ ☐

Q. 856. A weight is suspended from the middle of a rope whose ends are at the same level. In order for the rope to be perfectly horizontal, the forces applied to the ends of the rope

(*a*) must be greater than the weight ☐
(*b*) must be equal to the weight ☐
(*c*) might be so great as to break the rope ☐
(*d*) must be infinite. ☐

Q. 857. A hollow metal sphere is filled with water and is hung by a long thread. It is then made to oscillate. If water slowly flows through a small hole in the bottom, how will the period of oscillation be affected ?

(*a*) the period will go on decreasing till the sphere is empty ☐
(*b*) the period will go on increasing till the sphere is empty ☐
(*c*) the period will remain unchange throughout ☐
(*d*) the period will first increase, than it will decrease till it is empty and the period will be finally same as when the sphere was full of water. ☐

Q. 858. The minimum velocity of projection to go out from the earth's gravitational full is known as

(*a*) projectile velocity ☐
(*b*) escape velocity ☐
(*c*) angular velocity ☐
(*d*) terminal velocity. ☐

Q. 859. The square of a satellite time of revolution round the earth is

(*a*) directly proportional to the cube of radius of orbit ☐
(*b*) inversely proportional to the cube of radius or orbit ☐
(*c*) directly proportional to radius of orbit ☐
(*d*) inversely proportional to radius of orbit. ☐

Q. 860. The earth retains its atmosphere because

(*a*) the earth is sphere ☐
(*b*) the escape velocity is less than the mean speed of molecules ☐

(*c*) the earth has population ☐
(*d*) the escape velocity is more than the mean speed of molecules. ☐

Q. 861. To find time the astronaut in the earth satellite should use
(*a*) a pendulum clock ☐
(*b*) a watch having spring to keep it going ☐
(*c*) either of these two ☐
(*d*) None of the above. ☐

Q. 862. A body floats in a liquid contained in a beaker. The whole system as shown in figure falls freely under gravity. The up-thrust on the body due ot liquid is

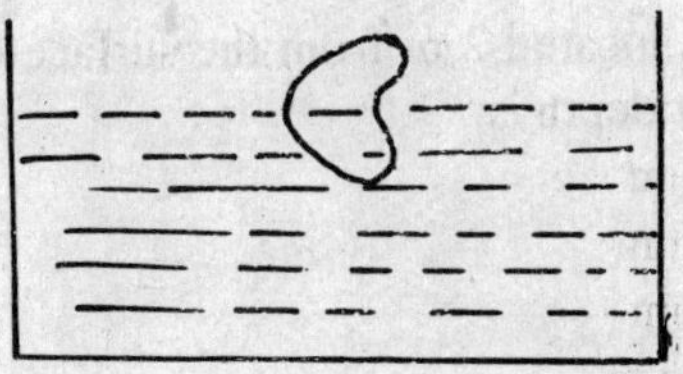

(*a*) zero ☐
(*b*) equal to the weight of the body in air ☐
(*c*) equal to the weight of the liquid displaced ☐
(*d*) equal to the weight of immersed portion of the body. ☐

Q. 863. The ratio of orbital velocity and escape velocity is
(*a*) $1 : \sqrt{2}$ ☐
(*b*) 2 :1 ☐
(*c*) $\sqrt{2} : 1$ ☐
(*d*) 4 : 1. ☐

Q. 864. Reynold number of two similar aeroplanes of different sizes
(*a*) is less in case of bigger aeroplane ☐
(*b*) is more is case of bigger aeroplane ☐
(*c*) is same for both ☐
(*d*) None of the above. ☐

Q. 865. Acceleration due to gravity is not affected by which one of the following ?

(*a*) Latitude ☐
(*b*) Altitude ☐
(*c*) Longitude ☐
(*d*) Depth. ☐

Q. 866. A rocket can go vertically upwards in earths atmosphere because

(*a*) of gravitational pull of the sun ☐
(*b*) it is higher than air ☐
(*c*) it has a fan which displaces more air per unit time than the weight of the rocket ☐
(*d*) of the force exerted on the rocket by gases ejected by it. ☐

Q. 867. A fish is located 2 m from the surface of a small pond. It's apparent depth is

(*a*) 1.0 m ☐
(*b*) 1.5 m ☐
(*c*) 2.0 m ☐
(*d*) 2.66 m. ☐

Q. 868. If the focal length of the eye lens in Huygen's eye-piece is f, then the focal length of the field lens and the separation between the two are respectively

(*a*) $3f$ and f ☐
(*b*) f and $\frac{3}{2}f$ ☐
(*c*) $3f$ and $2f$ ☐
(*d*) $2f$ and $3f$. ☐

Q. 869. If the incident rays are very oblique

(*a*) astimatism becomes predominant ☐
(*b*) spherical aberration becomes predominant ☐
(*c*) both astigmatism and spherical aberration appear ☐
(*d*) None of the above. ☐

Q. 870. In Frensel from of diffraction the distance between the screen and the obstacle/aperture is

(*a*) zero ☐
(*b*) infinite ☐

(*c*) finite ☐
(*d*) None of the above. ☐

Q. 871. For achromatic combination of lenses of if we use two lenses, having focal lengths f and f_1 and dispersive powers ω and ω respectively when

(*a*) $\omega = \omega_0,\ \omega' = 2\omega_0;\ f' = 2f$ ☐
(*b*) $\omega = \omega_0,\ \omega' = 2\omega_0;\ f' = -2f$ ☐
(*c*) $\omega = \omega_0,\ \omega' = 2\omega_0;\ f' = \frac{f}{2}$ ☐
(*d*) $\omega = \omega_0,\ \omega' = 2\omega_0;\ f' = -\frac{f}{2}$. ☐

Q. 872. The defect in image because of centric rays falling n the lens is known as

(*a*) astigmats ☐
(*b*) coma ☐
(*c*) curvature of the image field ☐
(*d*) spherical aberration. ☐

Q. 873. The function of field lens of an eye-piece is

(*a*) field lens with relatively large aperture helps increase the field of view for the observer ☐
(*b*) field lens in conjugation with eye lens reduces spherical aberration ☐
(*c*) Both (*a*) and (*b*) ☐
(*d*) eye-piece consists of only one lens. ☐

Q. 874. The distance between tow consecutive bright bonds in fringes is

(*a*) $\frac{D}{2d} \times \lambda$ ☐
(*b*) $\frac{2d}{d} \times \lambda$ ☐
(*c*) $\frac{2D}{\lambda} \times d$ ☐
(*d*) $2Dd\lambda$. ☐

Q. 875. In Newton's ring experiment the radius of nth dark ring for air is

(*a*) $\sqrt{nR\lambda}$ ☐
(*b*) $nR\lambda$ ☐

(c) $\frac{nR\lambda}{\mu}$ ☐

(d) $\frac{R\lambda}{n\mu}$. ☐

Q. 876. In case of Fraunhoffer diffraction the distance between source and diffracting screen

(a) zero ☐

(b) infinite ☐

(c) finite ☐

(d) None of these. ☐

Q. 877. The blurring of image because of dispersion in lens is known as

(a) spherical aberration ☐

(b) chromatic aberration ☐

(c) astigmatism ☐

(d) curvature image field. ☐

Q. 878. In case of Fraunhoffer diffraction the fringed image of formed is

(a) aperture ☐

(b) obstacle ☐

(c) source ☐

(d) All of these. ☐

Q. 879. The magnifying of a compound microscope in terms of magnification m_0, because of objective and the magnifying power mE because eye-piece is given by

(a) m_0/mE. ☐

(b) $m_0 \times mE$ ☐

(c) $m_0 + mE$ ☐

(d) $m_0 - mE$. ☐

Q. 880. The length of an astronomical telescope for normal vision is

(a) $f_0 \times f_e$ ☐

(b) f_0 / f_e ☐

(c) $f_0 + f_e$ ☐

(d) $f_0 - f_e$. ☐

Q. 881. A steel wire of length 'L' has a magnetic moment 'M'. It is then bent into a semi-ciruclar arc. The new magnetic moment is

(a) M

(b) $\frac{2M}{\pi}$

(c) $\frac{M}{\pi}$

(d) $M\pi$.

Q. 882. If a substance is placed in a skech glass on the pole pieces of a magnet, the substance accumulates on the sides causing a depression at the centre, it may be concluded that the substance is

(a) ferromagnetic in nature

(b) paramagnetic in nature

(c) diamagnetic in nature

(d) None of these.

Q. 883. A magnetic material which is prove to spontaneous magnetisation at sufficiently low temperature and whose magnetic moment due to electron spins in bulk are all oriented along an applied magnetic field are called

(a) ferromagnetic

(b) anti-ferromangnetic

(c) paramagnetic

(d) diamagnetic.

Q. 884. The volume susceptibility of a magnetic material is 30×10^{-4}. Then its relative permeability is

(a) 31×10^{-4}

(b) 29×10^{-4}

(c) 1.003

(d) 1.0003.

Q. 885. The area of hysteresis loop is proportional to the work done in

(a) magnetising of the substance

(b) demagnetising the substance

(c) reversing the magnetic field

(d) None of these.

Q. 886. Substances that experience a force from stronger to weaker regions of magnetic field are called

(a) non-magnetic

(*b*) paramagnetic ☐
(*c*) ferromagnetic ☐
(*d*) diamagnetic. ☐

Q. 887. Relative permeability is
(*a*) the ratio of flux density in a material to that for a vacuum ☐
(*b*) very large for paramagnetic materials ☐
(*c*) small for ferromagnetic materials ☐
(*d*) equal to $4\pi \times 10^{-7} T$. m/A for a vaccum. ☐

Q. 888. Lagging of magnetic flux density behind magnetic field is known as
(*a*) paramagnetism ☐
(*b*) diamagnetism ☐
(*c*) susceptibility ☐
(*d*) hysteresis. ☐

Q. 889. Van de Graff generators are sometimes built horizontally because they
(*a*) are so small ☐
(*b*) accelerate electrons ☐
(*c*) accelerate protons ☐
(*d*) None of these. ☐

Q. 890. The most energetic radiations are produced by
(*a*) linear accelerator ☐
(*b*) cosmic rays ☐
(*c*) cyclotron ☐
(*d*) synchrotron. ☐

Q. 891. In a synchrotron the magnetic field must change to compensate for
(*a*) heating of the coils ☐
(*b*) loss of energy due to air resistance ☐
(*c*) increases in the radius of circular path ☐
(*d*) None of these. ☐

Q. 892. Protons are accelerated in one step by which of the following?
(*a*) Van de Graff Generator ☐
(*b*) Cyclotron ☐

(c) Spark chamber ☐
(d) Linear accelerator. ☐

Q. 893. Electrons can be accelerated to very high energies by means of
(a) thyotrons ☐
(b) magnetrons ☐
(c) betatrons ☐
(d) cyclotrons. ☐

Q. 894. If a conductor is subjected to a magnetic field that is perpendicular to a current carrying conductor, a potential difference is observed between any two points on the opposite side, it is because of
(a) Hall effect ☐
(b) Peltier effect ☐
(c) Thomson effect ☐
(d) None of these. ☐

Q. 895. The relative weights of atom can be determined by which of the following ?
(a) Geiger counter ☐
(b) Wilson cloud chamber ☐
(c) Electron microscope ☐
(d) Mass spectrograph. ☐

Q. 896. Hall effect shows that majority of current carriers in a metal should be
(a) electrons ☐
(b) holes ☐
(c) positrons ☐
(d) None of these. ☐

Q. 897. A paramagnetic substance of a susceptibility 3×10^{-4} is placed in a magnetic field by 4×10^{4} ampere/meter. Then intensity of magnetisation is
(a) 1.33×10^{-8} ☐
(b) 0.75×10^{-8} ☐
(c) 12 ☐
(d) None of these. ☐

Q. 898. In rowland ring of annealed iron the winding has 100 turn/m when current in winding is 0.5 amperes, then H is

(*a*) 500 ☐
(*b*) 2000 ☐
(*c*) 5×10^{-4} ☐
(*d*) None of these. ☐

Q. 899. In case of a *P-N* junction diode at high value of reverse bias the current rises sharply. This value of reverse bias is called

(*a*) cut off voltage ☐
(*b*) zener voltage ☐
(*c*) transfer of power ☐
(*d*) transfer of voltage. ☐

Q. 900. Transistor stands for

(*a*) transfer for resistance ☐
(*b*) transfer of current ☐
(*c*) transfer of power ☐
(*d*) transfer of voltage. ☐

Section X

Q. 901. In a PN-junction diode, if junction current is zero, then

(*a*) the potential barrier has disappeared ☐
(*b*) no carriers are crossing the junction ☐
(*c*) equal number of majority ad minority carriers are crossing the junction ☐
(*d*) the number of holes that diffuse from *P* junction is equal to the number of electrons that diffuse from *N*-junction. ☐

Q. 902. In the active region operation of a transistor

(*a*) the collector and emitter junctions are forward biased ☐
(*b*) the collector and emitter junctions are reverse biased ☐
(*c*) the collector junction is forward biased and the emitter junction is forward biased ☐
(*d*) the collector junction is reverse biased and the emitter junction is forward biased. ☐

Q. 903. Why is a capacitor-resistor combination used in oscillators in grid circuit ?

(*a*) To generate oscillations ☐

(*b*) To bias the grid automatically ☐

(*c*) To amplify the oscillations ☐

(*d*) To sustain the oscillations. ☐

Q. 904. When we use a triode as class A amplifier, the grid bias should correspond to

(*a*) the straight line portion of V_z - I_p *graph* ☐

(*b*) the upper bend ☐

(*c*) the lower bend ☐

(*d*) the cut off grid voltage. ☐

Q. 905. A signal voltage is applied to the grid of a triode in a circuit containing plate load resistance, when the signal voltage applied to grid is at its maximum value ?

(*a*) the plate current is maximum ☐

(*b*) the plate current is minimum ☐

(*c*) the plate current is zero ☐

(*d*) the voltage on the plate must be at its maximum instantaneous value. ☐

Q. 906. Which one exerts the greatest influence on flow of electrons in a triode ?

(*a*) The supply voltage ☐

(*b*) Cathode ☐

(*c*) Anode ☐

(*d*) Grid. ☐

Q. 907. How is the amphncation factor affected in a triode if the grid is moved closer to the cathode ?

(*a*) It decreases ☐

(*b*) It incrcases ☐

(*c*) It may increase or decrease ☐

(*d*) It remains unaffected. ☐

Q. 908. What will be the effect on amplification factor in a triode if we space the grid wires wider ?

(*a*) It decreases ☐

(*b*) It increases ☐

(c) It may decrease or increase ☐
(d) It remains unaffected. ☐

Q. 909. What is the advantage of class *A* operation of an amplifier ?
(a) Its high efficiency ☐
(b) Its high gain ☐
(c) Its low distortion ☐
(d) None of these. ☐

Q. 910. What for is a modulator circuit used ?
(a) To eliminate carrier waves ☐
(b) To superpose the audio frequency signal on the carrier waves ☐
(c) To transmit the signal ☐
(d) None of these. ☐

Q. 911. What is the advantage of class *C* operation of an amplifier ?
(a) Its high efficiency ☐
(b) Its low distortion ☐
(c) Its high distortion ☐
(d) None of these. ☐

Q. 912. Which mode of operation of an amplifier is analogous to the operation of a rectifier ?
(a) Class *A* ☐
(b) Class *B* ☐
(c) Class *C* ☐
(d) Class *AB*. ☐

Q. 913. An oscillator is basically an amplifier with loop gain
(a) less than unity ☐
(b) more than unity ☐
(c) zero ☐
(d) None of these. ☐

Q. 914. Ripple factor in case of a half wave rectifier is equal to
(a) 1.1 ☐
(b) 1.21 ☐
(c) 0.482 ☐
(d) 1.482. ☐

Q. 915. Ripple factor in case of a full wave rectifier is equal to
(a) 1.1 ☐

(*b*) 1.21
(*c*) 0.482
(*d*) 1.482.

Q. 916. What is the name given to *N*-type part of the forward biased section in a *NPN* transistor ?
(*a*) Emitter
(*b*) Collector
(*c*) Base
(*d*) Donor.

Q. 917. What for are diodes used in power-supply circuits ?
(*a*) Rectifier
(*b*) Signal detector
(*c*) Wave shapers
(*d*) Amplifiers.

Q. 918. If 2 particles of masses m_1 and m_2 move under the action of interacting forces their refuced mass is
(*a*) $m_1 + m_2$
(*b*) $\frac{1}{m_1} + \frac{1}{m_2}$
(*c*) $\frac{m_1 m_2}{m_1+m_2}$
(*d*) None of these.

Q. 919. A curve track is banked for a certain velocity when a particle goes round the circular track with a lesser velocity. The force of friction perpendicular to the direction of motion
(*a*) is totally absent
(*b*) acts along the road outwards
(*c*) acts horizontally outwards
(*d*) acts along the road inwards.

Q. 920. If 2 equal forces have a resultant equal to the magnitude of either of the 2 forces then the angle between the 2 forces is
(*a*) 0°
(*b*) 60°
(*c*) 190°
(*d*) 120°.

Q. 921. If three forces acting on a rigid body satisfy Lame's theorem, then

(*a*) the body will be in equilibrium irrespective of whether the forces are coplaner or not ☐

(*b*) the body will be in equilibrium only when the forces are coplaner ☐

(*c*) the body will be in equilibrium only when the forces are not coplaner ☐

(*d*) the body will not be in equilibrium in either case. ☐

Q. 922. A lift falls freely under gravity. The vertical equilibrium of the body inside is

(*a*) stable ☐

(*b*) unstable ☐

(*c*) neutral ☐

(*d*) None of these. ☐

Q. 923. If a system having 3 movable weightless pulleys in which a separate string passes over each pulley, the mechanical advantage, assuming the system to be ideal, is

(*a*) 4 ☐

(*b*) 6 ☐

(*c*) 8 ☐

(*d*) 9. ☐

Q. 924. The coefficient of static friction between a body and the surface is 0.3 and the coefficient of kinetic friction between the body and the surface is 0.2. The weight of the body is 100 N and the surface is horizontal. A force of 10 Newtons acts on the body parallel to the surface. The functional force acting on the body is

(*a*) 10 N ☐

(*b*) 15 N ☐

(*c*) 20 N ☐

(*d*) 25 N. ☐

Q. 925. If the metacentric height of a floating body above its centre of gravity is zero, then on being slightly displaced from the equilibrium, it will

(*a*) remain in displaced position ☐

(*b*) return back to the original position ☐
(*c*) further displace to a new position ☐
(*d*) None of these. ☐

Q. 926. A solid body floats in a liquid of specific gravity 0.8 with 2/5th of its volume exposed to air. Its specific gravity is

(*a*) 0.6 ☐
(*b*) 0.48 ☐
(*c*) 0.4 ☐
(*d*) 0.32. ☐

Q. 927. The gravitational field at any point inside a spherical shell of radius *a* is

(*a*) GM/a^2 ☐
(*b*) GM/a ☐
(*c*) $-GM/a$ ☐
(*d*) zero. ☐

Q. 928. Yield point is the point at which

(*a*) a wire breaks ☐
(*b*) a wire elongates no longer ☐
(*c*) a wire has the extension increasing rapidly ☐
(*d*) None of these. ☐

Q. 929. Work done in producing a strain in unit volume of a substance is

(*a*) $\frac{1}{2} \times$ stress x strain ☐
(*b*) strain x strain ☐
(*c*) strain/stress ☐
(*d*) stress/strain. ☐

Q. 930. A beam is said to undergo uniform bending if

(*a*) it is placed symmetrically on two knife edges and loaded at the middle ☐
(*b*) it is placed symmetrically on two knife edges and loaded at 2 points equally, the 2 points being at the same distance on either side of the 2 knife edges ☐
(*c*) it is clamped at one end and loaded at the free end ☐
(*d*) None of the above. ☐

Q. 931. The angle of contact of water with glass is 0°. In a capillary tube, the meniscus of water will have the shape of a

(*a*) concave sphere ☐
(*b*) plane surface ☐
(*c*) hemisphere ☐
(*d*) None of the above. ☐

Q. 932. Due to surface tension, a tiny droplet of any liquid has the shape of

(*a*) a cylinder ☐
(*b*) a sphere ☐
(*c*) an ellipoid ☐
(*d*) None of the above. ☐

Q. 933. In the artificial rain making process, silver iodide released helps

(*a*) in further cooling the cloud ☐
(*b*) in acting as the condensation nuclei ☐
(*c*) in removing the dust particles from the cloud ☐
(*d*) not in any of the aforesaid methods. ☐

Q. 934. When the Kelvin temperature of a dilute solute of a non-electrolyte is doubled, its osmotic pressure becomes

(*a*) $\frac{1}{2}$ time ☐
(*b*) $\frac{1}{\sqrt{2}}$ time ☐
(*c*) $\sqrt{2}$ times ☐
(*d*) 2 times. ☐

Q. 935. Diffusion is a process in which

(*a*) molecules of a gas or liquid move from regions of higher concentration to regions of lower concentration only when aided by gravity ☐
(*b*) molecules of a gas or liquid move from regions of higher concentration to regions of lower concentration even it is against gravity ☐
(*c*) molecules of a gas or liquid move from regions of lower concentration only when aided by gravity ☐
(*d*) molecules of a gas or liquid move from regions of lower concentration to regions of higher concentration even if it is against gravity. ☐

Q. 936. Diffraction pattern of a straight edge is given by

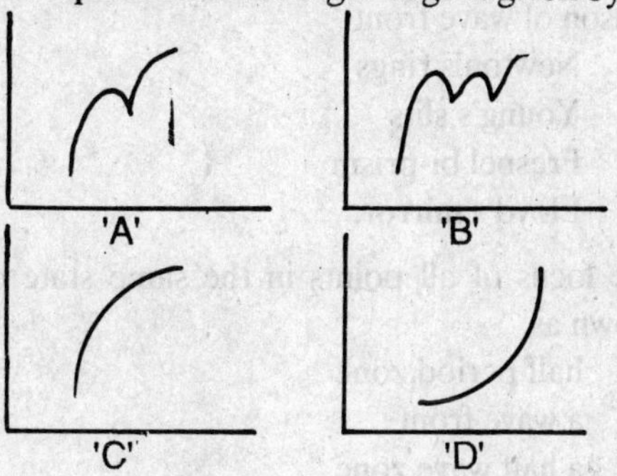

(*a*) *A* ☐
(*b*) *B* ☐
(*c*) *C* ☐
(*d*) *D*. ☐

Q. 937. In the non-localised fringes the distance of the *n*th bright band from the centre is given by

(*a*) $X_n = \frac{D}{d} \times n\lambda$ ☐

(*b*) $X_n = \frac{2D}{d} \times n\lambda$ ☐

(*c*) $X_n = \frac{D}{2} \times n\lambda$ ☐

(*d*) $X_n = \frac{D}{d} \times n\lambda^2$. ☐

Q. 938. The fringes in Newton's ring experiment are known as

(*a*) Faidingh fringes ☐
(*b*) Fringes of equal inclination ☐
(*c*) Fringes of equal thickness ☐
(*d*) Localised fringes. ☐

Q. 939. Maxwell based his theory of electromagnetic waves on the hypothesis that a changing electric field gives rise to

(*a*) an electric current ☐
(*b*) a stream of electrons ☐
(*c*) a magnetic field ☐
(*d*) longitudinal waves. ☐

Q. 940. Select the one that is not an example of interference by divison of wave front

(*a*) Newton's rings ☐
(*b*) Young's slits ☐
(*c*) Fresnel bi-prism ☐
(*d*) Lloyd's mirror. ☐

Q. 941. The locus of all points in the same state of vibrations are known as

(*a*) half period zone ☐
(*b*) a wave front ☐
(*c*) a half wave zone ☐
(*d*) None of these. ☐

Q. 942. Select the *Wrong* statement with reference to interference

(*a*) $X = \dfrac{Dn\lambda}{d}$, is a condition for a bright fringe ☐
(*b*) $X = \dfrac{(2n-1)\lambda}{2d}$, is a condition for a dark fringe ☐
(*c*) $B = \dfrac{D}{d} \times \dfrac{\lambda}{2}$, gives the fringe width of bright fringe ☐
(*d*) The distance between two successive dark and bright fringes is same. ☐

Q. 943. Which of the following colour phenomenon is not an interference phenomenon ?

(*a*) Oil spread on the surface of water exposed to extended source of light exhibits brilliant colours ☐
(*b*) Soap bubbles in sun light show colours ☐
(*c*) Sky appears blue at noon but red at evening and dawn ☐
(*d*) metallic surface when heated show colours. ☐

Q. 944. Longitudinal waves *cannot* be

(*a*) reflected ☐
(*b*) diffracted ☐
(*c*) dispersed ☐
(*d*) polarised. ☐

Q. 945. Images formed from real objects by diverging lenses are always

(*a*) virtual ☐

(b) enlarged ☐
(c) inverted ☐
(d) real. ☐

Q. 946. For a compound microscope, the image formed by the eye-piece is
(a) real ☐
(b) invertd ☐
(c) erect ☐
(d) diminished. ☐

Q. 947. For optical phenomenon, as described by electromagnetic theory, it is the electric vector which is involved, the magnetic vector introduction in theory is
(a) only customary ☐
(b) for representing the field energy or density flux ☐
(c) because with an electric vector, association is a must ☐
(d) for explaining anamolous dispersion. ☐

Q. 948. If two sources in an interference experiment are having wide slits, then
(a) maxima and minima will be close to each other ☐
(b) shadow will be formed ☐
(c) general illumination will be obtained ☐
(d) None of these. ☐

Q. 949. Conductors used for storing charge for a long time must be spherical in shape because
(a) charge is uniformly distributed over a spherical conductor ☐
(b) it can be easily charged ☐
(c) more surface is available for charge ☐
(d) it can be easily handled. ☐

Q. 950. Whenever an e.m.f. and current is set up by a charge of magnetic flux through a circuit its direction will be such as to oppose the act which caused it. This is known as
(a) Faraday's aw ☐
(b) Kirchhoff's law ☐
(c) Lenz's law ☐

(d) None of these. ☐

Q. 951. Electric charge
(a) resides on the outer surface of an insulated conductor ☐
(b) resides on the inner surface of an insulated conductor ☐
(c) resides on the outer as well as inner surfae of an insulated conductor ☐
(d) None of these. ☐

Q. 952. In case of Seebeck effect if A is neutral temperature and B is the temperature of inversion then
(a) $A = B$ ☐
(b) $B > A$ ☐
(c) $B < A$ ☐
(d) B may be less than or greater than A depending upon the condition of couple used. ☐

Q. 953. The space near a charged body within which its influence can be felt is called an electric
(a) region ☐
(b) atmosphere ☐
(c) field ☐
(d) All of these. ☐

Q. 954. A thermocouple is made of pure copper and pure iron. The cold junction is maintained at 0°C and hot junction heated from 40°C to 600°C. The thermoelectromotive force will
(a) rise to a maximum, decline to zero and then reverse direction ☐
(b) rise to a maximum and then decline to zero ☐
(c) rise to a maximum and then remain constant ☐
(d) rise steadily with temperature increase ☐

Q. 955. The relationship between capacity (C), charge (Q) and potential (V) is
(a) $C = \frac{Q}{V}$ ☐
(b) $C = QV$ ☐

(c) $C = \frac{V}{Q}$ ☐
(d) None of these. ☐

Q. 956. Of the following, which one is correct in respect to the Seebeck effect ?
(a) It is the same as Thomson effect ☐
(b) It is the inverse of Thamson effect ☐
(c) It is the same as the Peltier effect ☐
(d) It is the inverse of Peltier effect. ☐

Q. 957. Which of the following can be used to produce large potential difference ?
(a) Electrophorus ☐
(b) Wimburst machine ☐
(c) Gass plate machine ☐
(d) Van de Graff generator. ☐

Q. 958. What name is given to the phenomenon of emission of electrons from hot bodies ?
(a) Thermionic emission ☐
(b) Field emission ☐
(c) Photoelectric emission ☐
(d) Secondary emission. ☐

Q. 959. The majority carrier in n-type semiconductor is
(a) hole only ☐
(b) electron only ☐
(c) proton only ☐
(d) Both electron and hole. ☐

Q. 960. The amplification of triode depends upon
(a) filament voltage ☐
(b) plate voltage ☐
(c) plate current ☐
(d) plate resistance. ☐

Q. 961. In a pure semiconductor
(a) there is no conduction of electricity ☐
(b) electrical conduction is due to holes as well as electrons. ☐
(c) electrical conduction is due to holes only ☐

(*d*) electrical conduction is due to electrons only. ☐

Q. 962. Which of the following is a *correct* statement ?

(*a*) A diode can be used as a rectifier ☐

(*b*) A triode can be used as rectifier ☐

(*c*) The current in diode is alway proportional to the voltage ☐

(*d*) The curved portion of I_v V_s V_g is used for amplification of a triode. ☐

Q. 963. The majority carrier in a p-type semiconductor is

(*a*) electron only ☐

(*b*) proton ☐

(*c*) hole only ☐

(*d*) both electron and hole. ☐

Q. 964. In half-wave rectifier the r.m.s. value of a.c. component of the wave is

(*a*) more than d.c. valve ☐

(*b*) same as that of d.c. ☐

(*c*) less than d.c. valve ☐

(*d*) None of these. ☐

Q. 965. The amplifier circuit shown in figure below is

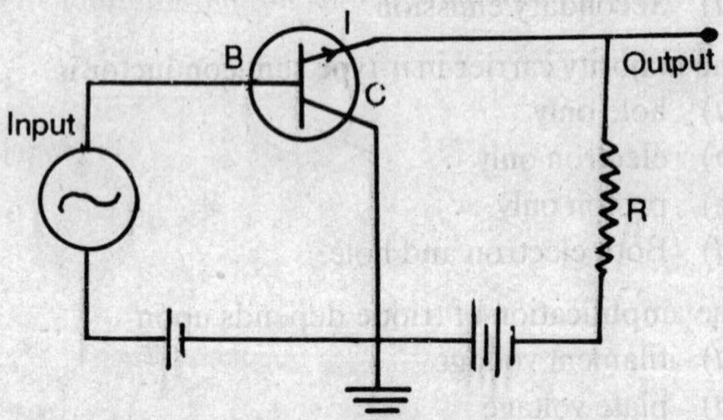

(*a*) a common emitter amplifier ☐

(*b*) a common base amplifier ☐

(*c*) a common collector amplifier ☐

(*d*) All of these. ☐

Q. 966. A semiconductor is the substance

(*a*) that contains a large number of free electrons ☐

(*b*) that forms covalent bonds with its neighbouring atoms ☐
(*c*) in which the number of free electrons is negligible ☐
(*d*) in which there is only one electron in the outermost orbit. ☐

Q. 967. What type of impurity is to be added to the semiconductor material to provide holes ?
(*a*) Monovalent ☐
(*b*) Trivalent ☐
(*c*) Tetravalent ☐
(*d*) Pentavalent. ☐

Q. 968. Choose the one that acts as signal inverter
(*a*) Common emitter ☐
(*b*) Common collector ☐
(*c*) Common base ☐
(*d*) None of these. ☐

Q. 969. Which of the following electronic devices is used for digital displays in electronic calculators ?
(*a*) Photodiodes ☐
(*b*) Zener diodes ☐
(*c*) LEDs ☐
(*d*) SRCs. ☐

Q. 970. The relation between thermionic current (*I*) and temperature (*T*) is
(*a*) $I \propto T$ ☐
(*b*) $I \propto T^2$ ☐
(*c*) $I \propto T^3$ ☐
(*d*) $I \propto \frac{1}{T^2}$. ☐

Q. 971. Which type of motion is significant in case of molecules of monoatomic gases ?
(*a*) Rotatory ☐
(*b*) Vibratory ☐
(*c*) Translatory ☐
(*d*) All of these. ☐

Q. 972. The r.m.s. of molecules in a sample of helium is 5/7th of that of the molecules of hydrogen. If the temperature of hydrogen sample is O°C, the sample of helium is

(*a*) 100°C ☐
(*b*) 273°C ☐
(*c*) O°C ☐
(*d*) O°K. ☐

Q. 973. The mass *M* of an ideal gas is filled in a closed container of volume *V* at a pressure *P* and absolute temperature *T*. An additional mass 2M is now introduced in the container and the volume is reduced to V/3 and temperature to T/3. The new pressure of gas will be

(*a*) 3*P* ☐
(*b*) 3 ☐
(*c*) 2*P* ☐
(*d*) *P*/3. ☐

Q. 974. Mean free path of gas molecules is inversely proportional to its

(*a*) volume ☐
(*b*) pressure ☐
(*c*) temperature ☐
(*d*) None of these. ☐

Q. 975. How many dead centres are there in one cycle of a steam engine ?

(*a*) 1 ☐
(*b*) 2 ☐
(*c*) 3 ☐
(*d*) 4. ☐

Q. 976. Bismuth expands on solidification. If pressure is applied to a sample of bismuth, it will

(*a*) decrease it melting point ☐
(*b*) not affect its melting point ☐
(*c*) increase its melting point ☐
(*d*) or will not effect its melting point. ☐

Q. 977. How many degrees of freedom does a butterfly moving in air possess

(*a*) 2 ☐
(*b*) 1 ☐
(*c*) 3 ☐
(*d*) depends upon the type of motion. ☐

Q. 978. The volume of a gas is held constant while its temperature is raised. The pressure the gas exerts on the walls of its container increases because

(*a*) each molecule loses more kinetic energy when it strikes the wall ☐
(*b*) the masses of the molecules increase ☐
(*c*) the molecules are in contact with the wall for a shorter time ☐
(*d*) the molecules have higher average speeds and strike the wall more often. ☐

Q. 979. Molecules are, in general apart from one another in

(*a*) gases ☐
(*b*) crystaline solids ☐
(*c*) liquids ☐
(*d*) amorphous solids. ☐

Q. 980. The ratio of Universal Gas Constant and Avogadro's number is called

(*a*) Boltzmann's Constant ☐
(*b*) Equilibrium Constant ☐
(*c*) Velocity Constant ☐
(*d*) None of these. ☐

Q. 981. At a given temperature, the average kinetic energy of gaseous bromine molecule

(*a*) is equal to ☐
(*b*) a little greater than ☐
(*c*) a little less than ☐
(*d*) can't be compared to the average kinetic energy of gaseous hydrogen molecules. ☐

Q. 982. The kinetic theory of matter

(*a*) is a tentative hypothesis yet to be confirmed by experiment ☐
(*b*) can be derived directly from the law of conservation of energy ☐

(*c*) contains assumption accepted because the resulting theory agrees with experimental data ☐
(*d*) cannot be considered valid because molecules cannot be readily seen or measured. ☐

Q. 983. The temperature at which oxygen molecules will have the same root mean square velocity as that of hydrogen molecules at – 100°C is
(*a*) 3040°C ☐
(*b*) 2768°C ☐
(*c*) 2500°C ☐
(*d*) 2495°C. ☐

Q. 984. Ship X is steaming due east at 12.0 knots, and ship Y is steaming due south at 16.0 knots. The speed of ship Y relative to ship X is
(*a*) 10.0 ☐
(*b*) 10.6 ☐
(*c*) 14.6 ☐
(*d*) 20.0. ☐

Q. 985. If *C* is proportional to *D*, then
(*a*) *C* is smaller than *D* ☐
(*b*) the ratio *C*/*D* is constant ☐
(*c*) the product *CD* is constant ☐
(*d*) *C* is larger than *D*. ☐

Q. 986. We know that forces are vectors because
(*a*) they obey the same mathematical rules that displacements obey ☐
(*b*) they are defined in such a way that they are vectors ☐
(*c*) they always add to zero ☐
(*d*) the have magnitude. ☐

Q. 987. An experiment involves the two measurements *a* and *b*. The result of the experiment is expressed as $Y = a \times b$. The maximum permissible error Δr in when the permissible error in the measurements of *a* and *b* are Δa and Δb respectively (*a*) $\left(\frac{\Delta r}{r}\right)_{max} = \frac{\Delta a}{a} + \frac{\Delta b}{b}$ ☐

(*b*) $\left(\frac{\Delta r}{r}\right)_{max} = (a/\Delta a) + b/\Delta b)$ ☐

(*c*) $\left(\frac{\Delta r}{r}\right)_{max} = (\Delta a)a + (\Delta b)c$ ☐

(*d*) None of these. ☐

Q. 988. The graph between two physical quantities is a straight line if the quantities are

(*a*) constants ☐
(*b*) equal in magnitude ☐
(*c*) directly proportional to each other ☐
(*d*) inversely proportional to each other. ☐

Q. 989. Huygen's conception of secondary waves

(*a*) is used to determine the velocity of light ☐
(*b*) is use to determine the wavelength of light ☐
(*c*) is a geometrical method to find a wave front ☐
(*d*) gives us magnifying power of a microscope. ☐

Q. 990. The necessary condition for production of sharp interference pattern is that frequency of light emitted by two sources should be

(*a*) 2 : 1 ☐
(*b*) 1 : 2 ☐
(*c*) same ☐
(*d*) different but close to each other. ☐

Q. 991. Select the one that is not a condition for permanent interference of light

(*a*) Two sources should continuously emit waves of the same wave length ☐
(*b*) Amplitude of two wave trains should be equal ☐
(*c*) The two sources should be very narrow ☐
(*d*) Interfering wave fronts must interfere at an angle of 45°. ☐

Q. 992. In a Bi-prism experiment if one slit is covered with black opaque paper

(*a*) no fringe will be obtained ☐
(*b*) the fringes will be narrower ☐
(*c*) the fringes will be broader ☐
(*d*) the fringes will be darker. ☐

Q. 993. The double slit experiment shown in figure below is used to estimate the wavelength of monochromatic light. If the path difference $S_2X - S_1X$ were known for a bright fringe at X, what other single measurement is required ?

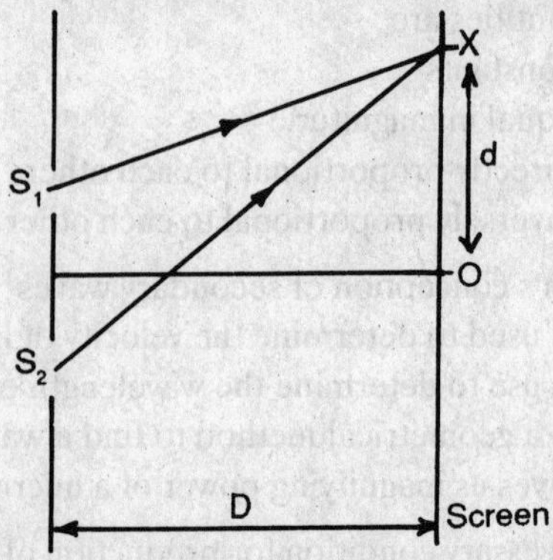

(*a*) The distance S_1S_2 between slits ☐
(*b*) The distance D from slit to screen ☐
(*c*) The distance d of the fringe from the axis ☐
(*d*) The number of bright fringes from O to X. ☐

Q. 994. To measure the distance between two coherent sources in a bi-prism experiment, the convex lens should be placed between

(*a*) the source and the slit ☐
(*b*) the slit and the Bi-prism ☐
(*c*) the Bi-prism and the eye-piece ☐
(*d*) None of these. ☐

Q. 995. The wave front is an imaginary surface at which

(*a*) constant phase difference is always maintained ☐
(*b*) phase is always same for all the points ☐
(*c*) phase changes with constant rate in all directions along the surface ☐

(*d*) phase changes with rate of $\frac{2\pi}{\lambda}$ per unit length in all directions along the surface. ☐

Q. 996. Interference can take palce between

(*a*) light waves only but not in waves on the surface of water ☐

(*b*) transverse waves only, but not in longitudinal waves ☐

(*c*) longitudinal waves only, but not in transverse waves ☐

(*d*) transverse waves as well as longitudinal waves. ☐

Q. 997. If two identical interfering waves produce a dark band of third order their path difference is

(*a*) zero ☐

(*b*) $\frac{7\lambda}{2}$ ☐

(*c*) $\frac{3\lambda}{2}$ ☐

(*d*) $\frac{5\lambda}{2}$. ☐

Q. 998. One of the tubes of Jamin's interferometer contains air at room temperature and atmospheric pressure, while the removal of air from the other corresponds to a shift of 150 fringes. If the wave length of light be 5×10^{5} cms and the tybe are 25 cms long, the refractive index of air is

(*a*) 1.003 ☐

(*b*) 1.0003 ☐

(*c*) 100.3 ☐

(*d*) 0.10003. ☐

Q. 999. If one of the two slits in a Bi-prism experiment is painted over so that it transmits half the light intensity of the other

(*a*) the dark fringes will be brighter and the brighter fringes darker ☐

(*b*) the bright fringes and dark fringes both will be dark ☐

(*c*) the fringes system will disappear altogether ☐

(*d*) the bright fringes will be brighter and dark fringes darker. ☐

Q. 1000. Resultant amplitude at a point due to the entire wave front is

(*a*) $R_1 + R_2 + R_3 + ...$ ☐

(*b*) R_1 ☐

(*c*) $\frac{R_1}{2}$ ☐

(*d*) None of these. ☐

Section XI

Q. 1001. Which of the following phenomenon can not be understood by quantum theory ?

(*a*) Photoelectric effect ☐

(*b*) Compton effect ☐

(*c*) X-ray production ☐

(*d*) Interferenc. ☐

Q. 1002. Diffraction pattern is observed on a screen due to a circular disc illuminated by a plane wave front. If size of circular disc is increased then intensity at centre

(*a*) increases ☐

(*b*) decreases ☐

(*c*) remains same ☐

(*d*) Cannot be predicted. ☐

Q. 1003. Which one of the following relation is incorrect ? The symbols have their usual meanings

(*a*) $C_p - C_v = R \frac{\text{joules}}{\text{mole °K}}$ ☐

(*b*) $W = HJ$ ☐

(*c*) $C_p - C_v = \frac{R}{J}$ cal/deg $C_p . C_v$ where $J = 3.18 \times 10^7$ ergs/mole °K ☐

(*d*) $C_p . C_v = R$. ☐

Q. 1004. In the operation of a heat engine heat is at the particular temperature and

(*a*) is partly converted into mechanical energy and partly exhausted at the same temperature ☐
(*b*) is partly converted into mechanical energy and partly exhausted at a lower temperature ☐
(*c*) is completely converted into mechanical energy ☐
(*d*) is partly converted into mechanical energy and partly exhausted at a higher temperature. ☐

Q. 1005. A heat engine can develop efficiency equal to 100% if the temperature of the sink is

(*a*) less than that of the source ☐
(*b*) equal to that of the source ☐
(*c*) O°K ☐
(*d*) O°C. ☐

Q. 1006. During an adiabatic gas expansion, the environment

(*a*) serves as a heat sink ☐
(*b*) serves as a heat source ☐
(*c*) must be at a higher temperature than the gas ☐
(*d*) does not have to participate. ☐

Q. 1007. The stem engines A and B have their sources at 900°K and 600°K and their sinks at 450°K and 300°K respectively

(*a*) They are equally efficient ☐
(*b*) A is less efficient than B ☐
(*c*) A is more efficient than B. ☐
(*d*) Their efficiencies cannot be determined ☐

Q. 1008. If the temperature of the source and sink are increased by same amount, the efficiency of the engine

(*a*) decreases ☐
(*b*) remains unchanged ☐
(*c*) increases ☐
(*d*) may increase or decrease. ☐

Q. 1009. Which of the following should not change in isothermal operation ?

(*a*) heat content ☐
(*b*) volume ☐

(c) temperatuer ☐
(d) pressure. ☐

Q. 1010. Heat cannot be transferred from a colder to a hotter body uless work is done. This is a statement of

(a) First law of thermodynamics ☐
(b) Efficiency of a heat engine ☐
(c) Second law of Thermodynamics ☐
(d) Adiabatic process. ☐

Q. 1011. The ideal thermal efficiency of a cyclic heat engine is limited by

(a) the friction in the engine ☐
(b) the amount in the engine ☐
(c) the difference between input temperature and output temperature ☐
(d) the amount of work. ☐

Q. 1012. All natural processes

(a) decrease random motion of atoms ☐
(b) are reversible ☐
(c) increase entropy ☐
(d) decrease entropy. ☐

Q. 1013. The internal energy of a monoatomic gas is

(a) 3/2 RT ☐
(b) independent of temperature ☐
(c) in the form of kinetic energy ☐
(d) partially kinetic and partially potential. ☐

Q. 1014. The temperature of inversion is

(a) $T_i = 2arb$ ☐

(b) $T_i = \frac{Rb}{2a}$ ☐

(c) $T_i = \frac{2a}{Rb}$ ☐

(d) $T_i = \frac{2Ra}{b}$. ☐

Q. 1015. On moving up in earth's atmosphere the magnitude of mean free path of air molecules

(a) increases ☐

(*b*) remains constant ☐
(*c*) decreases ☐
(*d*) cannot be predicted. ☐

Q. 1016. The raio of C_p and C_v can be found with the help of
(*a*) Regnault's method ☐
(*b*) Callender and Barnes continuous flow method ☐
(*c*) Andrews method ☐
(*d*) Celement and Desorme's method. ☐

Q. 1017. All natural processes proceed in the direction of
(*a*) increased temperature ☐
(*b*) decreased order ☐
(*c*) increased order ☐
(*d*) decreased entropy. ☐

Q. 1018. A bullet moving with a uniform velocities *v* stop suddenly after hitting the target and the whole mass melts. If the mass of the bullet be m gm, specific heat is initial temperature 25°C, melting point 475°C and latent heat *L*, *v* is then given by

(*a*) $ms\,(475\text{-}25) + mR = \frac{2J}{mv^2}$ ☐

(*b*) $ms\,(475\text{-}25) + mL = \frac{1}{2}\frac{mv^2}{J}$ ☐

(*c*) $mL = ms\,(475\text{-}25) + \frac{1}{2}mv^2\frac{1}{J}$ ☐

(*d*) $ms\,(475\text{-}25) = mL = \frac{2J}{mv^2}$. ☐

Q. 1019. Whenever work is transformed intoheat or vice-versa the quantity of work is mechanically equivalent to the quantity of heat is statement of
(*a*) Newton's law ☐
(*b*) Joule's law ☐
(*c*) First law of Thermodynamics ☐
(*d*) Second law of Thermodynamics. ☐

Q. 1020. The state of the given mass of an encolsed gas can be completely specified by knowing
(*a*) three thermodynamics variables ☐

(*b*) two thermodynamics variables
(*c*) one thermodynamic variables
(*d*) None of above.

Q. 1021. What makes the air coming out of a punctured tyre cool ?
(*a*) Adiabatic expansion
(*b*) Flow at high speed
(*c*) Isothermal expansion
(*d*) None of these.

Q. 1022. What type of the process is the Carnot's cycle ?
(*a*) Reversible
(*b*) Neither reversible nor irreversible
(*c*) Irreversible
(*d*) May be reversible or irreversible.

Q. 1023. Which of the following is *not* true for images formed by a plane mirror ?
(*a*) magnification is + 1
(*b*) Image distance is negative
(*c*) Right and left are reversed
(*d*) Images are real.

Q. 1024. The index of refraction for a substance is
(*a*) constant
(*b*) constant for a given wavelength
(*c*) variable with the speed of light
(*d*) never constant.

Q. 1025. Total internal reflection occurs when the angle of incidence is
(*a*) greater than the angle of refraction
(*b*) equal to the critical angle
(*c*) greater than the critical angle
(*d*) greater than 45°.

Q. 1026. A negative magnification always means that the image is
(*a*) erect
(*b*) real
(*c*) virtual
(*d*) inverted.

Q. 1027. Which of the following is *not* true for images formed by real objects located inside the focal point of a converging lens ?

(a) Virtual ☐
(b) Erect ☐
(c) Real ☐
(d) enlarged. ☐

Q. 1028. Which of the following demonstrates the transverse nature of lightwaves ?
(a) Interference ☐
(b) Polarization ☐
(c) Diffraction ☐
(d) Refraction. ☐

Q. 1029. The wavelength of light from a distant object does not affect
(a) interference ☐
(b) resolving power ☐
(c) diffraction ☐
(d) polarization. ☐

Q. 1030. A diffraction grating with a spacing of 15,000 lines/in has a slit separation of
(a) 6.67×10^{-6} in ☐
(b) 5.9×10^{-3} cm ☐
(c) 3231 nm ☐
(d) 1.69×10^{-4} cm. ☐

Q. 1031. Ultrasonic sound waves
(a) can be heard by a normal human ear ☐
(b) can be heard with the help of hearing apparatus ☐
(c) cannot be heard ☐
(d) can be heard with the help of microphone. ☐

Q. 1032. If a car is moving towards you with its horn sounding, you hear a higher pitch than you would hear if the car were at rest. This is because
(a) the waves travel faster ☐
(b) successive crests produced by the horn are closer to each other ☐
(c) you recived successive crests more frequently even though the wave itself is unchanged ☐
(d) the motion of the car raises the frequency of the horn. ☐

Q. 1033. The quality of a tone

(*a*) decreases with loudness ☐
(*b*) varies directly as pitch ☐
(*c*) varies inversely as its frequency ☐
(*d*) depends upon the overtones that are present there. ☐

Q. 1034. Two particles are executing S. H. M. of same period, *T*. If the second particle starts from its mean position $T/2$ later than the first, the phase difference between the two particles at any instant of time is

(*a*) $\pi/2$ ☐
(*b*) π ☐
(*c*) $2\pi/2$ ☐
(*d*) 2π. ☐

Q. 1035. The amplitude of sound wave determines its

(*a*) loudness ☐
(*b*) pitch ☐
(*c*) reverberation ☐
(*d*) interference. ☐

Q. 1036. Overtones are

(*a*) boats ☐
(*b*) the fundamental notes produced in pipe ☐
(*c*) the notes produced in a pipe other than the fundamental ☐
(*d*) all of the above. ☐

Q. 1037. When a tuning fork is made to vibrate, vibrations of its two prongs

(*a*) are in phase ☐
(*b*) have a phase difference of π ☐
(*c*) have a phase difference $\pi/2$ ☐
(*d*) have a phase difference of $\pi/4$. ☐

Q. 1038. The velocity of the wave could be increased by

(*a*) reducing the amplitude ☐
(*b*) decreasing the frequency ☐
(*c*) increasing the period ☐
(*d*) stretching the spring more. ☐

Q. 1039. A circle is obtained as a Lassajour figure when simple harmonic motions are having

(*a*) the same amplitude and a phase difference of π ☐

(*b*) the same amplitude and a phase difference of $\pi/2$ ☐

(*c*) the same amplitude, frequency and a phase difference of π ☐

(*d*) the same amplitude, frequency and a phase difference of $\pi/2$. ☐

Q. 1040. Neglecting damping, the amplitude of the wave

(*a*) is the same throughout ☐

(*b*) varies periodically down the length of the spring ☐

(*c*) drops off inversely with distance ☐

(*d*) drops off as the square of the distance. ☐

Q. 1041. If E is the coefficient of volume elasticity of the medium, ρ its density, then the velocity of a longitudinal wave in fluid is given by

(*a*) $v = E/\rho$ ☐

(*b*) $v = \sqrt{E/\rho}$ ☐

(*c*) $v = \rho/E$ ☐

(*d*) $v = \sqrt{\rho/E}$. ☐

Q. 1042. Velocity of sound as given by Laplace is

(*a*) $v = \sqrt{\gamma P/\rho}$ ☐

(*b*) $v = \rho\sqrt{\gamma/\rho}$ ☐

(*c*) $v = P\sqrt{\gamma/\rho}$ ☐

(*d*) $v = \gamma\sqrt{P/\rho}$. ☐

Q. 1043. The magnetic flux linked with a circuit when a unit current flows through it is knwon as

(*a*) induced current ☐

(*b*) induced e.m.f. ☐

(*c*) coeficient of self induction of a circuit ☐

(*d*) None of these. ☐

Q. 1044. The steady current which sends the same charge through a circuit in the same time as the alternating current does in half its period is known as

(*a*) induced current ☐

(*b*) steady current ☐

(*c*) mean value of alternating current ☐
(*d*) None of these. ☐

Q. 1045. For measuring very high temperature we use
(*a*) Mercury thermometer ☐
(*b*) Platinum resistance thermometer ☐
(*c*) Thermoelectric pyrometer ☐
(*d*) None of the above. ☐

Q. 1046. That steady current which produces the same heating effect in a resistance in a given time as the alternating current does in the same resistance in the same time is called
(*a*) induced current ☐
(*b*) root mean square value of an alternating current ☐
(*c*) mean value of alternating current ☐
(*d*) None of the above. ☐

Q. 1047. The effect in which e.m.f. is produced by a thermocouple when one of the junctions is heated keeping the other cold is called
(*a*) Peltier effect ☐
(*b*) Seebeck effect ☐
(*c*) Joule effect ☐
(*d*) Thomson effect. ☐

Q. 1048. The resistance due to inductance air an A.C. circuit is known as
(*a*) reactance ☐
(*b*) self-inductance ☐
(*c*) mutual inductance ☐
(*d*) None of these. ☐

Q. 1049. The effective resistance in an A.C. circuit is an
(*a*) inductance ☐
(*b*) impedence ☐
(*c*) mutual inductance ☐
(*d*) None of these. ☐

Q. 1050. The absorption or evolution of heat energy if a current flows along a conductor when different parts of the conductor are at different temperatures is known as

(*a*) Seebeck's effect ☐
(*b*) Peltier's effect ☐
(*c*) Thomson's effect ☐
(*d*) None of these. ☐

Q. 1051. Rate of change of thermo e.m.f. with temperature is known as

(*a*) Peltier's coefficient ☐
(*b*) Thomson's coefficient ☐
(*c*) Thermoelectric power ☐
(*d*) None of the above. ☐

Q. 1052. Reciprocal of impedence is called

(*a*) inductance ☐
(*b*) capacitance ☐
(*c*) admittance ☐
(*d*) None of these. ☐

Q. 1053. If the conductor is fixed and the slowly drifting charged particles, constituting the current, be displaced sideways with in the conductor under the action of a transverse magnetic field, it should be possible to find out the number of charged particles this is known as

(*a*) Fradays effect ☐
(*b*) Hall's effect ☐
(*c*) Joule Thomson effect ☐
(*d*) None of these. ☐

Q. 1054. The Hall electric field strength (potential gradient/unit magnetic field/per unit current density is called

(*a*) Hall current density ☐
(*b*) Hall voltage ☐
(*c*) Hall coefficient ☐
(*d*) None of these. ☐

Q. 1055. The arrangements used to impart high energy to the atomic particles are known as

(*a*) generators ☐
(*b*) motors ☐
(*c*) particle accelerators ☐
(*d*) None of these. ☐

Q. 1056. An apparatus for accelerating particles (atomic) even by using moderate voltage is called

(*a*) generator ☐
(*b*) cyclotron ☐
(*c*) motor ☐
(*d*) None of these. ☐

Q. 1057. The charges + q are placed at the four corners of a square. How much charge must be placed at its centre so that the whole system is in equilibrium ?

(*a*) $-\frac{q}{4}(1+2\sqrt{2})$ ☐

(*b*) $-\frac{q}{2}(1+2\sqrt{2})$ ☐

(*c*) $-\frac{q}{8}(1+2\sqrt{2})$ ☐

(*d*) $+\frac{q}{4}(1-2\sqrt{2})$. ☐

Q. 1058. Three equal charges + q are placed on the circumference of a circle such that they form an equilateral triangle. The intensity of electric field at the centre of the circle is

(*a*) zero ☐

(*b*) $\frac{1}{4\pi\varepsilon_0} \cdot \frac{3q}{r^2}$ ☐

(*c*) $\frac{1}{4\pi\varepsilon_0} \cdot \frac{q}{r^2}$ ☐

(*d*) $\frac{1}{4\pi\varepsilon_0} \cdot \frac{3q}{r}$. ☐

Q. 1059. A certain physical quantity is calculated from the formula $(\pi/3)\ h\ (a^2 - b^2)$, where h, and b are all lengths. The quantity being calculated is *a*

(*a*) length ☐
(*b*) velocity ☐
(*c*) area ☐
(*d*) volume. ☐

Q. 1060. If the two vectors $\vec{A}$ and $\vec{B}$ are such that their cross-product A × B = O, vectors are

(*a*) inclined at an angle of 60 ☐
(*b*) parallel to each other ☐
(*c*) perpendicular to each other ☐
(*d*) inclined at an angle of 45. ☐

Q. 1061. Scientific theories

(*a*) must constantly be reviewed to see whether they are in accordance with new experimental observations ☐
(*b*) are summaries of particular experiments ☐
(*c*) represent guesses that have not yet been compared with observational data. ☐
(*d*) are laws of nature that are not subject to revision. ☐

Q. 1062. Which of the following pairs of physical quantities have the same dimensions ?

(*a*) Momentum and impulse ☐
(*b*) Energy and surface energy ☐
(*c*) Momentum and angular momentum ☐
(*d*) Force and surface tension. ☐

Q. 1063. One big advantage of the atomic standards of length and time is that

(*a*) they are of human size ☐
(*b*) they can be measured with simple equipment ☐
(*c*) they are natural units ☐
(*d*) they can be duplicated anywhere. ☐

Q. 1064. When a bomb dropped from an aeroplane explodes in air. Its total

(*a*) momentum decreases ☐
(*b*) momentum increases ☐
(*c*) kinetic energy increases ☐
(*d*) kinetic energy decreases. ☐

Q. 1065. The M.K.S. unit of energy is

(*a*) Dyne/sec ☐
(*b*) Joule ☐
(*c*) None ☐
(*d*) Dyne/sec. ☐

Q. 1066. In S.I. system Pascal is a unit of

(*a*) Surface tension ☐
(*b*) Surface ☐
(*c*) Viscosity ☐
(*d*) Velocity. ☐

Q. 1067. Forces of 3N, 4N and 12N act at a point in mutually perpendicular direction. The magnitude of the resultant force in newtons is:

(*a*) 13 ☐
(*b*) 11 ☐
(*c*) 5 ☐
(*d*) indeterminate form the information given. ☐

Q. 1068. When a body undergoes an acceleration

(*a*) its velocity increases ☐
(*b*) its speed increases ☐
(*c*) it falls toward the earth ☐
(*d*) a force acts upon it. ☐

Q. 1069. The minimum of unequal forces whose vector sum can equal to zero is

(*a*) 2 ☐
(*b*) 1 ☐
(*c*) 3 ☐
(*d*) 4. ☐

Q. 1070. An automobile that is towing a trailer is accelerating on a level road. The force that the automobile exerts on the trailer is

(*a*) equal to the force the trailer on the road ☐
(*b*) greater than force the trailer exerts on the automobile ☐
(*c*) equal to the force the trailer exerts on the automobile ☐
(*d*) equal to the force the road exerts on the trailer. ☐

Q. 1071. A force of 3N acts perpendicularly to a force of 4N. Their resultant has a magnitude of

(*a*) 7 N ☐
(*b*) 5 N ☐

(*c*) 1 N

(*d*) 12 N.

Q. 1072. Which of the following represents curve for current voltage for a photo cell ?

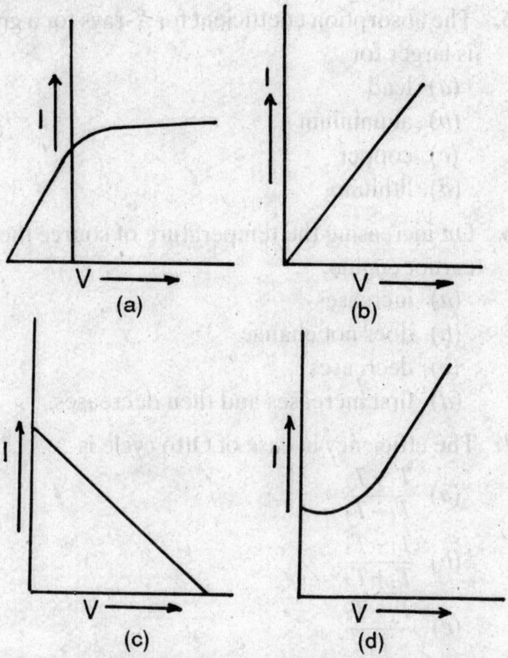

Q. 1073. The Bohr's formula when used for calculating ionisation energy of He, higher values than experimental values are obtained. It is due to

(*a*) electron-electron interaction

(*b*) spin-orbit interaction

(*c*) spin-spin interaction

(*d*) None of these.

Q. 1074. Broglie wavelength of an electron accelerated to a potential difference V volts is given by

(*a*) $\frac{150}{V}$ Å

(*b*) $\frac{h}{\sqrt{2mE}}$

(c) $\frac{h}{\sqrt{2mKT}}$ ☐

(d) R $\frac{150}{meV}$Å. ☐

Q. 1075. The absorption coefficient for X-rays for a given wavelength is larger for

(a) lead ☐
(b) aluminium ☐
(c) copper ☐
(d) lithium. ☐

Q. 1076. On increasing the temperature of source the efficiency of a carnot engine

(a) increases ☐
(b) does not change ☐
(c) decreases ☐
(d) first increases and then decreases. ☐

Q. 1077. The efficiency in case of Otto cycle is

(a) $\frac{T_1+T_2}{T_1-T_2}$ ☐

(b) $\frac{T_1-T_2}{T_1+T_2}$ ☐

(c) $\frac{T_1-T_2}{T_1}$ ☐

(d) $\frac{T_1 T_2}{T_1+T_2}$. ☐

Q. 1078. The efficiency of a Diesel engine with a compression ratio P, is given by

(a) $P\gamma - 1$ ☐

(b) $\left(\frac{1}{P}\right)^{\gamma-1}$ ☐

(c) $1 + P\gamma - 1$ ☐

(d) $1 - \left(\frac{1}{P}\right)^{\gamma-1}$ ☐

Q. 1079. The direction of heat flow is always from

(a) high pressure to low pressure ☐

(*b*) high temperature to low temperature ☐
(*c*) high density to low density ☐
(*d*) a point of higher emissivity. ☐

Q. 1080. Electromagnetic radiation is emitted
(*a*) only by bodies at higher temperatures to their surroundings ☐
(*b*) only by radio and television antennas ☐
(*c*) only by bodies at lower temperatures than their surrounding ☐
(*d*) by all bodies. ☐

Q. 1081. The amount of water contained in the air of a given room is most accurately described by the
(*a*) absolute humidity ☐
(*b*) vapour pressure ☐
(*c*) relative humidity ☐
(*d*) dew point. ☐

Q. 1082. Four identical copper cubes are painted with different types of paint. Which one would you expect to lose heat most rapidly, if all cubes are heated to the same temperature and are left in vacuum ?
(*a*) Painted shiny white ☐
(*b*) Painted rought white ☐
(*c*) Painted rought black ☐
(*d*) Painted shiny black. ☐

Q. 1083. The flow of heat from a hot to a cold body is an example of
(*a*) adiabatic process ☐
(*b*) irreversible process ☐
(*c*) reversible process ☐
(*d*) isothermal process. ☐

Q. 1084. A system of molecules whose energies are distributed in the most probable way
(*a*) can perform an amount of mechanical work ☐
(*b*) can perform an amount of mechanical work equal to its total energy content ☐
(*c*) cannot perform any mechenical work ☐
(*d*) None of the above. ☐

Q. 1085. A heat engine with 100% of efficiency would have to

(*a*) do no work ☐
(*b*) be at a uniform temperature ☐
(*c*) use no heat ☐
(*d*) discharge at 0°C. ☐

Q. 1086. On increasing the temperature of the source and keeping the temperature of the engine

(*a*) remains unchanged ☐
(*b*) decreases ☐
(*c*) increses ☐
(*d*) may increase or decrease. ☐

Q. 1087. The process in which no heat enters or leaves the system is termed as

(*a*) isochoric ☐
(*b*) isobaric ☐
(*c*) isothermal ☐
(*d*) adiabatic. ☐

Q. 1088. The ratio between the energy dissipated in some process and the heat that appears as a result is the

(*a*) specific heat ☐
(*b*) triple point ☐
(*c*) kilocalorie ☐
(*d*) mechanical equivalent of heat. ☐

Q. 1089. Conservation of energy means that

(*a*) energy can be destroyed but not created ☐
(*b*) energy can be created but not destroyed ☐
(*c*) energy can be both created and destroyed ☐
(*d*) energy can be neither created nor destroyed. ☐

Q. 1090. Whenever a system is made to complete a cyclic process, the work done during the complete cycle is

(*a*) is zero ☐
(*b*) positive ☐
(*c*) is negative ☐
(*d*) depends upon the path followed. ☐

Q. 1091. Two Reversible engines *A* and *B* have their sources at 800 K and 640 K. Their sinks are at 400K and 300K respectively

(*a*) *A* is more efficient than *B* ☐

(*b*) *A* is less efficient than *B* ☐
(*c*) They are equally efficient ☐
(*d*) Their efficiencies depend only upon the type of fuels used. ☐

Q. 1092. In an A.C. circuit maximum voltage is 423 volts, its effective voltage is
(*a*) 400 V ☐
(*b*) 300 V ☐
(*c*) 323 V ☐
(*d*) 340 V. ☐

Q. 1093. The substances have been classified ferromagnetic, paramagnetic and diamagnetic. It is according to
(*a*) molecular theory ☐
(*b*) electronic theory ☐
(*c*) both the above theories ☐
(*d*) None of these. ☐

Q. 1094. The supply voltage of A.C. mains is 240 V. The peak value of voltage is
(*a*) 300 V ☐
(*b*) 332.36 V ☐
(*c*) 336.32 V ☐
(*d*) 330.32 V. ☐

Q. 1095. Which of the following will have the same dimensions as those of frequency ?
(*a*) $\frac{1}{RC}$ ☐
(*b*) $\frac{R}{L}$ ☐
(*c*) $\frac{1}{LC}$ ☐
(*d*) $\frac{C}{L}$. ☐

Q. 1096. In an A.C. circuit, the current
(*a*) is in phase with voltage ☐
(*b*) leads the voltage ☐
(*c*) lags the voltage ☐
(*d*) any of the above depending upon the circumstances. ☐

Q. 1097. If the magnetic field in a cyclotron is halved and the radius of does in quadrupled, then the kinetic energy increases by a factor of

(*a*) 4 ☐
(*b*) 8 ☐
(*c*) 2 ☐
(*d*) zero. ☐

Q. 1098. Synchro-cyclotron removes which of the following limitations of a cyclotron ?

(*a*) Magnet's cost limitations ☐
(*b*) Magnets size limitations ☐
(*c*) Relativistic mass limitations ☐
(*d*) None of these. ☐

Q. 1099. Van de Graff generator is used for

(*a*) producing heat ☐
(*b*) generating charged particles ☐
(*c*) producing high voltage of the order of mega volts. ☐
(*d*) None of these. ☐

Q. 1100. *dV* is the change in potential between neighbouring points *dr* apart. The electric field E is then given by relation

(*a*) $E = -\dfrac{dV}{dr}$ ☐

(*b*) $E = dV \, . \, dr$ ☐

(*c*) $E = \dfrac{dr}{dV}$ ☐

(*d*) $E = \dfrac{dV}{dr}$. ☐

Section XII

Q. 1101. An A.C. power supply connected to an oscilloscope give the trace shown in Figure below

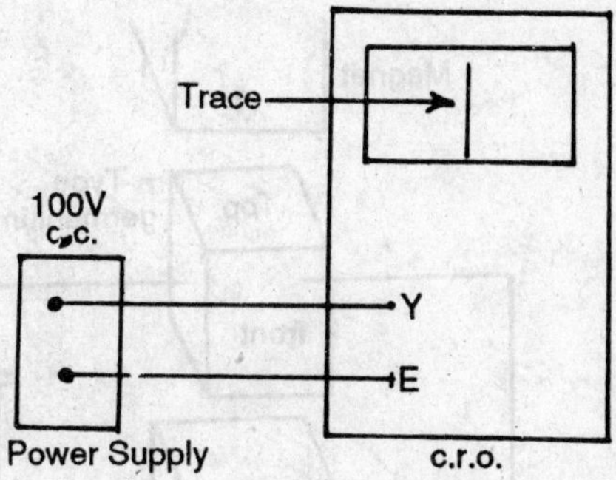

What adjustment has to be made to the oscilloscope so that the frequency of the A.C. supply can be measured ?

(*a*) Interchange leads to *Y* and *E* ☐
(*b*) Adjust *Y* gain control ☐
(*c*) Adjust focus control ☐
(*d*) Adjust time-base control. ☐

Q. 1102. What is the difference between a *p*-type germanium crystal and an ordinary germanium crystal ?

(*a*) The *p*-type crystal is positively charged due to a deficiency of electrons ☐
(*b*) The *p*-type crystal is negatively charged due to a surplus of electrons ☐
(*c*) The *p*-type crystal contains impurity atoms of higher atomic number than germanium ☐
(*d*) The *p*-type crystal contains impurity atoms of lower atomic number than germanium. ☐

Q. 1103. A cube of *p*-type germanium is placed between the poles of a magnet and a voltage is applied across opposite faces of the cube as shown in Figure

What effect will occur at the surface of the cube ?

(*a*) The top surface of the cube will become negatively charged ☐
(*b*) The front surface of the cube will become positively charged ☐
(*c*) The front surface of the cube will become negatively charged ☐

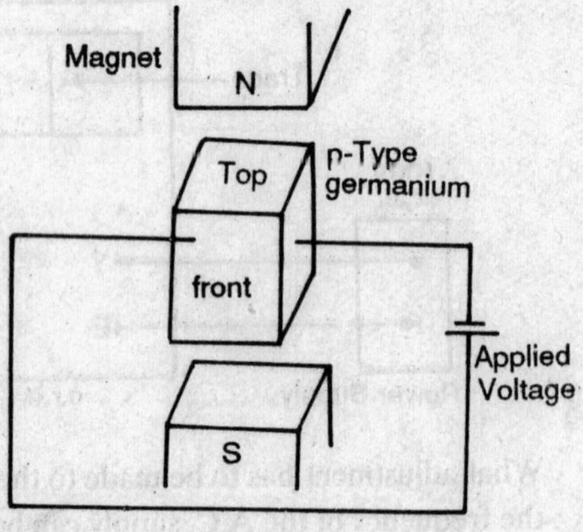

(*d*) Both top and front surface of the cube will become positively charged. ☐

Q. 1104. In class *A* amplifiers, current load flows for

(*a*) full-cycle of the signal ☐

(*b*) half-cycle of the signal ☐

(*c*) one-fourth cycle of the signal ☐

(*d*) All are true. ☐

Q. 1105. Ripple factor in case of a half-wave rectifer is equal to

(*a*) 1.1 ☐

(*b*) 1.21 ☐

(*c*) 0.482 ☐

(*d*) 1.482. ☐

Q. 1106. Ripple factor in case of full-wave rectifier is equal to

(*a*) 1.1 ☐

(*b*) 1.21 ☐

(*c*) 0.482 ☐

(*d*) 1.482. ☐

Q. 1107. The practical result of Schottky defect is that it

(*a*) increases the barrier height at the metal surface ☐

(*b*) decreases the barrier height at the metal surface ☐

(*c*) holds the barrier steady at metal surface ☐

(*d*) None of these. ☐

Q. 1108. A typical acceptor atom in *p*-type germanium has how many valence electrons ?

(*a*) 1 ☐
(*b*) 3 ☐
(*c*) 4 ☐
(*d*) 5. ☐

Q. 1109. *PN*-junction can be used as

(*a*) amplifier ☐
(*b*) oscillator ☐
(*c*) modulator ☐
(*d*) rectifier. ☐

Q. 1110. The potential barrier at a *PN*-junction is because of the charge of either side of the junction. These charges are

(*a*) minority carriers ☐
(*b*) majority carries ☐
(*c*) both majority and minority carrier ☐
(*d*) fixed donor and acceptor ions. ☐

Q. 1111. Why is there junction current at equilibrium in an unbiased *PN*-junction ?

(*a*) Because of diffusion of minority carriers only ☐
(*b*) Because of diffusion of majority carriers only ☐
(*c*) Zero, because equal and opposite carriers are crossing the junction ☐
(*d*) Zero, because no charges are crossing the junction. ☐

Q. 1112. An *X*-ray beam striking a thin gold foil may cause

(*a*) the formation of unstable nuclei ☐
(*b*) the increase of atomic number target atom ☐
(*c*) electrons to leave the gold foil ☐
(*d*) None of these. ☐

Q. 1113. From which of the following the space group is determined ?

(*a*) continuous *X*-ray spectra ☐
(*b*) characteristic absent spectra of *X*-rays ☐
(*c*) Moseley law ☐
(*d*) None of these. ☐

Q. 1114. Which of the following will not produce a track of condensed vapour in cloud chamber ?

(*a*) Electron ☐
(*b*) Proton ☐
(*c*) Positron ☐
(*d*) Neutron. ☐

Q. 1115. The value of Bohr's magneton is given by

(*a*) $\frac{eh}{4\pi c}$ ☐
(*b*) $\frac{eh}{4\pi mc}$ ☐
(*c*) $\frac{h}{4\pi m}$ ☐
(*d*) $\frac{eh}{4\pi mc}$. ☐

Q. 1116. For a given quantum number '*n*', the possible number of Sommerfield's elliptical orbits are

(*a*) one ☐
(*b*) zero ☐
(*c*) $n + 1$ ☐
(*d*) n. ☐

Q. 1117. The emission of an electron by an atom without the emission of X or γ-radiations as a result of change from an excited state to a lower state is known as

(*a*) Anger effect ☐
(*b*) Bohr's correspondence principle ☐
(*c*) Ritz's combination principle ☐
(*d*) isomerism. ☐

Q. 1118. The effect of Larmor precession is to

(*a*) increase the energy of the atom when excited ☐
(*b*) increase or decrease the energy of the atom when excited ☐
(*c*) decrease the energy of the atom when excited ☐
(*d*) maintain constant energy of the atom when excited. ☐

Q. 1119. In an electron microscope, as the concentration of the electron beam striking the object increases, the perceptibility

(*a*) decreases ☐
(*b*) increases ☐
(*c*) remains unchanged ☐
(*d*) first increases, levels off and then decreases exponentially. ☐

Q. 1120. In the Bohr model of the atom, the electron does not fall into the nucleus as

(*a*) it is not allowed by quantum rules ☐
(*b*) the electrostatic attraction is balanced by mechanical forces ☐
(*c*) the electron is not a particle ☐
(*d*) the space between the nucleus and the atom boundry is filled with ether. ☐

Q. 1121. Nuclear force as compared with electrostatic force is

(*a*) weaker and long range ☐
(*b*) weaker and short range ☐
(*c*) stronger and long range ☐
(*d*) stronger and short range. ☐

Q.1122. According to the law of parity

(*a*) every particle has a strangeness number ☐
(*b*) energy is conserved in nuclear reaction ☐
(*c*) every reaction has mirror image reaction ☐
(*d*) symmetry is possible. ☐

Q. 1123. The lose of mass in a.m.u. in a nuclear change can be calculated from energy involved in MeV by

(*a*) dividing by 931 ☐
(*b*) multiplying by 931 ☐
(*c*) dividing by 391 ☐
(*d*) multiplying by 391. ☐

Q. 1124. In the carbon cycle from which stars hotter than sun obtain their, energy, the ${}_6C^{12}$ isotope

(*a*) splits into three alpha particles ☐
(*b*) fuses with another ${}_6C^{12}$ nucleus to form ${}_{12}Mg^{24}$ ☐
(*c*) is completely converted into energy ☐
(*d*) is regenerated at the end of the cycle. ☐

Q. 1125. An orbiting electron is likely to lose energy in the form of

(*a*) radiation ☐
(*b*) conduction ☐
(*c*) convection ☐
(*d*) All of these. ☐

Q. 1126. Who introduced the concept that an atom possess quantised energy levels ?

(*a*) Raman ☐
(*b*) Bohr ☐
(*c*) Newton ☐
(*d*) Fermi. ☐

Q. 1127. In which of the following an electron emits energy ?

(*a*) When it is moving in any specified orbit ☐
(*b*) when it jumps from higher energy level to lower energy level ☐
(*c*) When it goes from lower energy level to higher energy level ☐
(*d*) None of these. ☐

Q. 1128. In accordance with Bohr's first postulate the kinetic energy ($\frac{1}{2}m^2V$) of the electron is equal to

(*a*) $\frac{1}{2} \cdot \frac{Ze^2}{r}$ ☐
(*b*) $\frac{Ze^2}{r}$ ☐
(*c*) $\frac{Ze^2}{r^2}$ ☐
(*d*) $\frac{1}{2} \cdot \frac{Ze^2}{r^2}$. ☐

Q. 1129. Lines of Balmer series are emitted by hydrogen atom when the electron jumps

(*a*) from the first orbit to any other orbit ☐
(*b*) from higher orbits to second orbit ☐
(*c*) from highr orbits to first orbit ☐
(*d*) from the third orbit to higher orbits. ☐

Q. 1130. The angular speed of an electron in the *n*th orbit of Bohr's hydrogen atom is

(a) directly proportional to n ☐
(b) directly proportional to n^2 ☐
(c) inversely proportional to n^2 ☐
(d) inversely proportional to n. ☐

Q. 1131. N identical spherical drops charged to the same potential V are combined to form a big drop. The potential of new drop wil be

(a) V ☐
(b) V/N ☐
(c) $V \times N$ ☐
(d) $V \times N^{2/3}$. ☐

Q. 1132. Ferromagnetic substances have

(a) high permeability and low susceptibility ☐
(b) high susceptibility and low permeability ☐
(c) high permeability and high susceptibility ☐
(d) low permeability and low susceptibility. ☐

Q. 1133. An oil drop is found floating freely between the plates of a parallel plate condenser, the plates being horizontal and the lower plate carrying a charge $+Q$. The area of each plate is A and the distance of separation between them is D. The mass of oil drop is M. In C.G.S. units the charge of the drop must be (g is acceleration due to gravity)

(a) $(A/Q)\ (g/M)$ ☐
(b) $MgA/4\pi Q$ ☐
(c) $-\frac{gA}{D} Q$ ☐
(d) $-\frac{MAg}{4\pi Q}$. ☐

Q. 1134. If the permeability of material is 0.9998, then it is

(a) diamagnetic ☐
(b) paramagnetic ☐
(c) ferromagnetic ☐
(d) non-magnetic. ☐

Q. 1135. A high-resistance A.C. voltmeter and a c.r.o. are connectd across resistor R as shown in Figure below. The Y amplitude control of the c.r.o. is set at 20 V/cm

What is the reading on the voltmeter ?

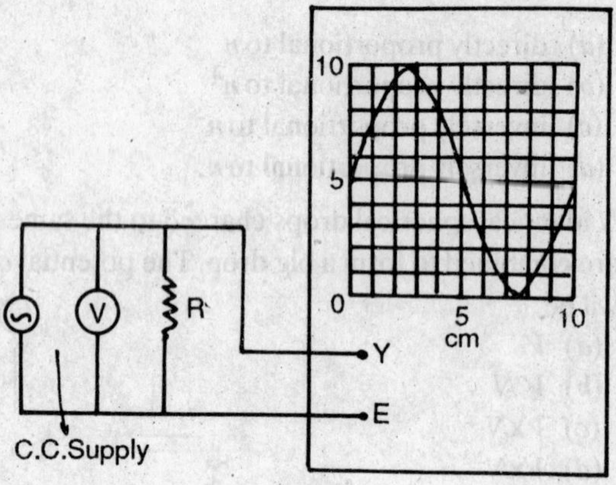

(*a*) 71 V
(*b*) 100 V
(*c*) 200 V
(*d*) 284 V.

Q. 1136. The switch *S* in the circuit in Figure below is open

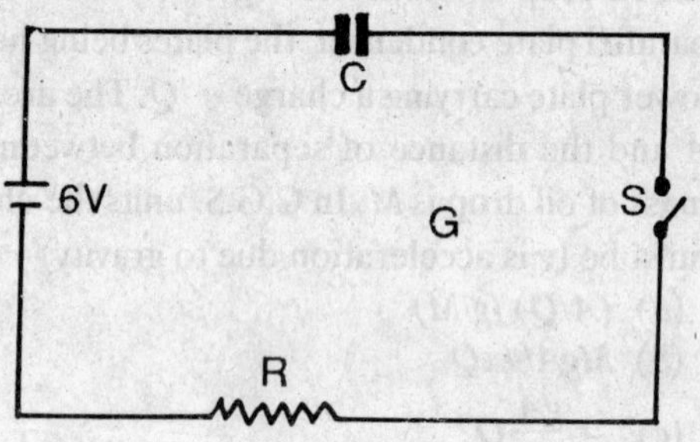

Which graph in Figure shows how the current varries with time, from the instant when *S* is closed ?

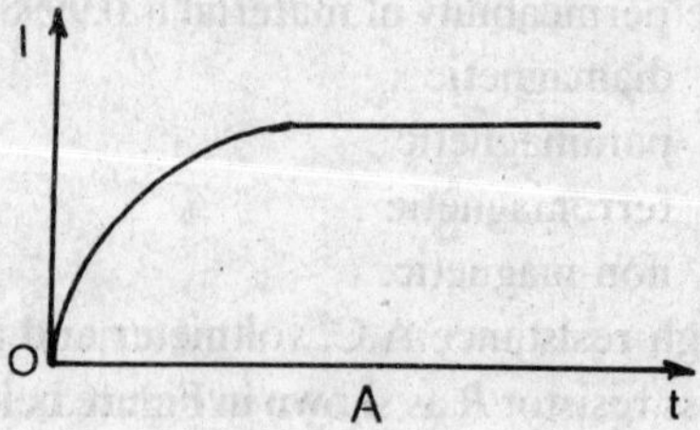

A

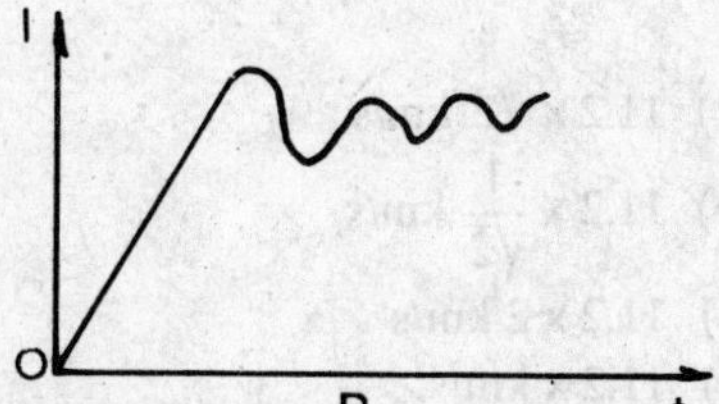

B

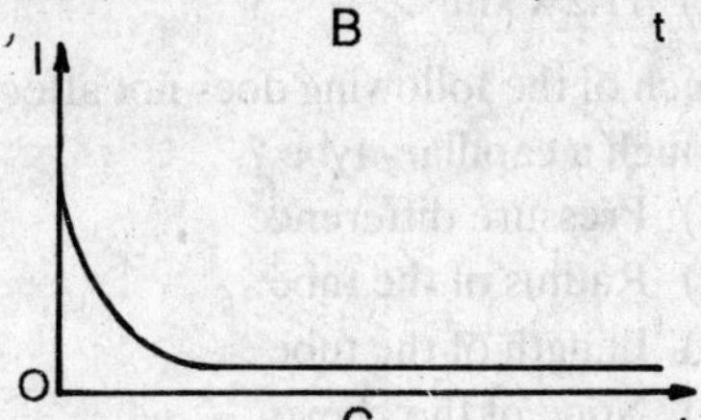

C

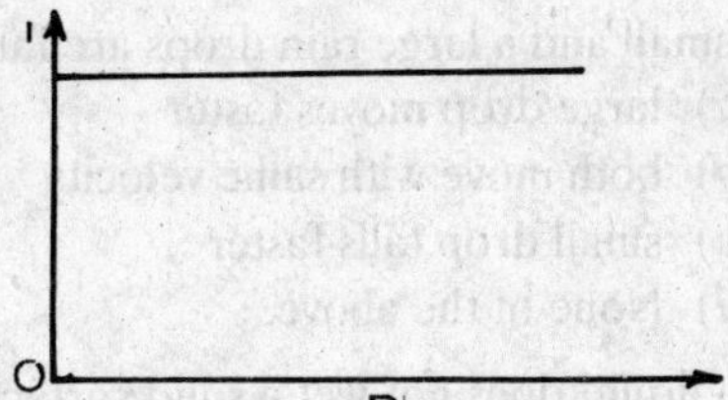

D

(*a*) A ☐
(*b*) B ☐
(*c*) C ☐
(*d*) D. ☐

Q. 1137. A projectile is fired horizontally with an initial speed of 20 m/s. Its horizontal speed 3s later is

(*a*) 20 m/s ☐
(*b*) 6.67 m/s ☐
(*c*) 60 m/s ☐
(*d*) 29.4 m/s. ☐

Q. 1138. The vertical speed of the above projectle after 3s is approximately

(*a*) 9.8 m/s ☐
(*b*) 60 m/s ☐
(*c*) 29.6 m/s ☐
(*d*) 20 m/s. ☐

Q. 1139. The escape velocity for a body projected vertically upwards from the surface of the earth is 11.2 km/s. If the body is projected in a direction making an angle of 45° with the vertical, the escape velocity will be:

(*a*) $11.2 \times \sqrt{2}$ km/s ☐
(*b*) $11.2 \times \frac{1}{\sqrt{2}}$ km/s ☐
(*c*) 11.2×2 km/s ☐
(*d*) $11.2 \times$ km/s. ☐

Q. 1140. Which of the following does not affect the flow of the liquid through a capillary tybe ?
(*a*) Pressure difference ☐
(*b*) Radius of the tube ☐
(*c*) Length of the tube ☐
(*d*) None of the above. ☐

Q. 1141. A small and a large rain drops are falling through air
(*a*) large drop moves faster ☐
(*b*) both move with same velocity ☐
(*c*) small drop falls faster ☐
(*d*) None of the above. ☐

Q. 1142. If a liquid does not wet a solid surface, the angle of contact is
(*a*) less than 90° ☐
(*b*) greater than 180 ° ☐
(*c*) 90° ☐
(*d*) between 90° and 180°. ☐

Q. 1143. Viscosity in fluids refers to
(*a*) the density of a fluid ☐
(*b*) tangential forces exerted on solid surface by the flowing liquid ☐
(*c*) normal forces exerted on solid surfaces by the flowing fluid ☐
(*d*) the compressibility of a fluid. ☐

Q. 1144. How does viscosity affect the relative motion between the liquid layers ?
(*a*) Does not affect ☐
(*b*) Accelerates ☐
(*c*) May accelerator or retard ☐
(*d*) Retards. ☐

Q. 1145. The shape of the mercury meniscus in glass container indicates that the

(*a*) cohesion between mercury molecules is greater than their adhesion to glass ☐
(*b*) cohesion between mercury molecules is equal to their adhesion to glass ☐
(*c*) cohesion between mercury molecules is greater than the cohesion of glass molecules ☐
(*d*) cohesion between glass molecules is equal to the cohesion between mercury molecules. ☐

Q. 1146. Two glass plates having a little water in between cannot be easily separated because of

(*a*) viscosity ☐
(*b*) atmospheric pressure ☐
(*c*) surface tension ☐
(*d*) friction. ☐

Q. 1147. The product of velocity and cross-sectional area for a liquid flowing through a pipe is a measure of the

(*a*) rate of flow ☐
(*b*) volume of fluid ☐
(*c*) fluid pressure ☐
(*d*) None of these. ☐

Q. 1148. Turbulent flow occurs when the Rynold number is above

(*a*) 1000 ☐
(*b*) 500 ☐
(*c*) 3000 ☐
(*d*) 5000. ☐

Q. 1149. The pressuse at the bottom of a tank of liquid is not proportional to

(*a*) the density of the liquid ☐
(*b*) the area of the liquid surface ☐
(*c*) the height of the liquid ☐
(*d*) the acceleration. ☐

Q. 1150. The hydraulic press operates primarily on

(*a*) Boyle's law ☐
(*b*) Pascal's law ☐

(c) Archimedes principle ☐
(d) Newton's laws. ☐

Q. 1151. What is the shape of the liquid meniscus in a capillary tybe placed in a liquid that wets the surface of the tybe ?
(a) Flat ☐
(b) Concave downwards ☐
(c) Convex upwards ☐
(d) None of the above is correct. ☐

Q. 1152. The shape of the water meniscus in a glass container indicates that
(a) the cohesion of water to glass is greater than the cohesion between water molecules ☐
(b) the cohesion between water molecules is greater than their adhesion to glass ☐
(c) the adhesion of water to glass is greater than the cohesion between water molecules ☐
(d) the cohesion between the glass molecules is equal to the cohesion between water molecules. ☐

Q. 1153. A good rocket fuel must be
(a) readily combustible ☐
(b) in gaseous state ☐
(c) capable of producing amounts of energy ☐
(d) capable of producing large volumes of gases. ☐

Q. 1154. A rocket propulsion is based on the principle of
(a) conservation of mass ☐
(b) conservation of energy ☐
(c) conservation of momentum ☐
(d) conservation of velocity. ☐

Q. 1155. For an electromagnet
(a) the retentivity of material should be large ☐
(b) the susceptibility of the material should be small ☐
(c) the retentivity small and susceptibility large ☐
(d) None of these. ☐

Q. 1156. To explain fine structure of the spectrum of hydrogen atom which of the following must be taken into consideration ?
(a) Orbital angular momentum ☐

(*b*) Spin angular momentum ☐
(*c*) The presence of neutrons in the nucleus ☐
(*d*) Finite size of the nucleus. ☐

Q. 1157. A hydrogen atom is in *p*-state. What are the values of *J* for this ?

(*a*) – 1/2, + 1/2, + 3/2 ☐
(*b*) – 1/2, – 3/2 ☐
(*c*) 3/2, 1/2 ☐
(*d*) 5/2, 3/2, 1/2. ☐

Q. 1158. What is the ratio of energies of hydrogen atom in its first and second excited states ?

(*a*) 1/4 ☐
(*b*) 2/3 ☐
(*c*) 4/9 ☐
(*d*) 4. ☐

Q. 1159. Which of the following was proved by Frank-Hertz experiment ?

(*a*) Electrons can have only certain values of energy ☐
(*b*) Atoms can accept energy only in certain discrete amount ☐
(*c*) The energy of an electron can be measured ☐
(*d*) The energy of the electron is the same as that of an emitted photon. ☐

Q. 1160. The existence of sharp lines in the spectrum of a gas indicates that

(*a*) electrons orbit at certain frequencies only ☐
(*b*) electrons emit energy in small indivisible packets of certain definite sizes ☐
(*c*) the nucleus has certain discrete energy levels ☐
(*d*) light exists in photons. ☐

Q. 1161. What is the radius of first Bohr's orbit for Z = 5 ?

(*a*) 0.53A ☐
(*b*) 0.106A ☐
(*c*) 2.65A ☐
(*d*) None of these. ☐

Q. 1162. Radioactive substances do not emit

(*a*) α-ray
(*b*) β-rays
(*c*) γ-rays
(*d*) protons.

Q. 1163. According to Bohr's theory of hydrogen atom the velocity of electron in the orbit of radius *r* is

(*a*) directly proportional to *r*
(*b*) inversely proportional to *r*
(*c*) directlly poportional to square root of *r*
(*d*) directly proportional to square of *r*.

Q. 1164. According to Bohr's theory the radius of quantized orbit is given by

(*a*) $\frac{4\pi^2 m}{n^2h^2Zc^2}$
(*b*) $\frac{n^2h^2}{4\pi^2 mZe^2}$
(*c*) $\frac{4\pi^2 mZe^2}{n^2h^2}$
(*d*) $\frac{n^2h^2Zc^2}{4\pi^2 m}$.

Q. 1165. What is the speed of an electron moving about in a circular orbit ?

(*a*) $\frac{mr^2}{Zc^2}$
(*b*) $\frac{Ze^2}{Zc^2}$
(*c*) $\frac{M\,.\,Ze^2}{mr}$
(*d*) $\frac{mr^2}{Ze^2}$.

Q. 1166. When an electron in an atom of hydrogen jumps from an outer orbit $n = 3$ to an inner orbit $n = 2$, the wavelength or emitted radiations will be equal to ?

(*a*) $\lambda = \frac{R}{6^2}$

(*b*) $\lambda = \frac{6^2}{R}$ ☐

(*c*) $\lambda = \frac{5R}{36}$ ☐

(*d*) $\lambda = \frac{36}{5R}$. ☐

Q. 1167. In the spectrum of which of the following will you find Balmer series ?

(*a*) Oxygen ☐

(*b*) Nitrogen ☐

(*c*) Hydrogen ☐

(*d*) All of these. ☐

Q. 1168. Which of the following can exist only in integral multiples of $h2/\pi$, according to assumption of Bohr's theory ?

(*a*) Principal quantum number ☐

(*b*) Linear momentum ☐

(*c*) Angular momentum ☐

(*d*) Azumuthal quantum number. ☐

Q. 1169. Which of the following relation gives the possible values by angular momentum for one valence electron ?

(*a*) $\sqrt{6} \cdot \frac{h}{4\pi}$ or $\sqrt{35} \cdot \frac{4h}{4\pi}$ ☐

(*b*) $\sqrt{35} \cdot \frac{h}{4\pi}$ or $\sqrt{63} \cdot \frac{h}{4\pi}$ ☐

(*c*) $\sqrt{12} \cdot \frac{h}{4\pi}$ or $\sqrt{20} \cdot \frac{h}{4\pi}$ ☐

(*d*) $\sqrt{15} \cdot \frac{h}{4\pi}$ or $\sqrt{6} \cdot \frac{h}{4\pi}$. ☐

Q. 1170. Bernoulli's equation includes as a special case

(*a*) Newton's third law of motion ☐

(*b*) Hooke's law ☐

(*c*) Torricelli's theorem ☐

(*d*) Archimede's Principle. ☐

Q. 1171. Elasticity of a gas under isothermal conditions is equal to its

(*a*) volume ☐

(*b*) pressure ☐

(*c*) force ☐

(*d*) None of the above. ☐

Q. 1172. The moment of inertia of a body comes in action :

(*a*) in circular motion ☐
(*b*) in straight line motion ☐
(*c*) in curved path ☐
(*d*) None of the above. ☐

Q. 1173. The force of friction that comes into action after the motion has started is known as:

(*a*) Dynamic friction ☐
(*b*) Static friction ☐
(*c*) Friction only ☐
(*d*) Limiting friction. ☐

Q. 1174. If the earth suddenly loses its attraction, then the mass of a body lying on its surface

(*a*) will remain same but weight will be no more ☐
(*b*) will become zero but its weight will be same ☐
(*c*) will become zero and weight also will be zero ☐
(*d*) will change. ☐

Q. 1175. The path traced out by a pebble emerging out of running motor-car's tyre is; (Neglect air resistance)

(*a*) Parabolic ☐
(*b*) Hyperbolic ☐
(*c*) Semi Parabolic ☐
(*d*) Cycloidal. ☐

Q. 1176. A motor boat is sailing in a ocean with a water line mark on it. Now the same motor boat sails in fresh water lake. The line mark would now be

(*a*) below the surface of water ☐
(*b*) above the surface of water ☐
(*c*) at the same line ☐
(*d*) remarkably below the surface. ☐

Q. 1177. There is some load in a pan attached with the spring. The pan is pulled down with some force and released. It start its up and down movements. the K.E. of that load will be :

(*a*) greatest in the middle of its movement ☐
(*b*) greatest at its upper extreme position ☐

(c) greatest at its lower extreme position ☐
(d) same in all the positions. ☐

Q. 1178. The theory of relativity shows that Newtonian mechanics is
(a) approximately correct only for velocities much smaller than the velocity of light ☐
(b) approximately correct for all velocities ☐
(c) wholly correct ☐
(d) correct only for velocities up to the velocity of light. ☐

Q. 1179. If the acceleration due to gravity be represented by unity, and one second unit of time, the unit of length is
(a) 981 cm ☐
(b) 100 cm ☐
(c) 98 cm/8^2 ☐
(d) 981 cm/s. ☐

Q. 1180. If the fundamental units are the velocity of light (3×10^8 m/s) acceleration due to gravity (10/m/s^2) and the mass of the proton, the unit of time in the system is
(a) 3×10^9 seconds ☐
(b) 3×10^8 seconds ☐
(c) 3×17^7 seconds ☐
(d) 10 seconds. ☐

Q. 1181. One Pascal in SI is equal to
(a) one newton metre^{-2} ☐
(b) one newton metre^{-1} ☐
(c) one newton metre^{-2} ☐
(d) one dyne cm^2. ☐

Q. 1182. A quantitative explanation of Brownian motion was given by
(a) Albert Einstein ☐
(b) Maxwell ☐
(c) Robert Brown ☐
(d) Boltzmann. ☐

Q. 1183. Point out *wrong* statement:
(a) Brownian motion helps the determination of the Avogadro number ☐

(*b*) Brownian motion supports the kinetic theory ☐

(*c*) In Brownian motion, the suspended particles show random motion ☐

(*d*) Brown first saw the molecules of the liquid in random motion. ☐

Q. 1184. Point out the *wrong* statement:

(*a*) In a virial expansion for gas, the second virial coefficient is zero at the Boyle temperature ☐

(*b*) There is no such thing as Boyle temperature for a perfect gas ☐

(*c*) the first virial coefficient is the same for all gases ☐

(*d*) above the Boyle temperature, the second virial coefficient is negative. ☐

Q. 1185. The term 'gas' denotes a substance in the gaseous state at a temperature

(*a*) above the critical temperature (T_o) ☐

(*b*) above the Boyle temperature ☐

(*c*) below the critical temperature ☐

(*d*) below the temperature of inversion. ☐

Q. 1186. Certain gases are called permanent gases because

(*a*) they cannot be liquefied ☐

(*b*) they are perfect gases ☐

(*c*) their critical temperatures are very low ☐

(*d*) their boiling point is low. ☐

Q. 1187. In the liquefaction of air carbon dioxide must be removed from the air because

(*a*) its presence lowers the temperature of inversion ☐

(*b*) it is an impurity not required ☐

(*c*) its presence reduces the rate of cooling ☐

(*d*) it will solidify and clog the system. ☐

Q. 1188. For a given difference fo pressure the Joule Kelvin cooling

(*a*) decreases with the fall in initial temperature ☐

(*b*) increases with the decrease of initial temperature ☐

(*c*) does not depend on initial temperature ☐

(*d*) first decreases and than increases. ☐

Q. 1189. In the Claude's air liquefier, the gas is first cooled by doing external work, adiabatically, because

(*a*) its parts do not require any lubrication ☐
(*b*) it is easier to arrange for this expansion ☐
(*c*) for the same difference of pressure the drop in temperature is greater than Joule Thomson Cooling when the initial temperature is high ☐
(*d*) for the same difference of temperature the drop in temperature is always greater then Joule Thomson Cooling. ☐

Q. 1190. Th lowest temperature that can be reached by boiling Helium under reduced pressure is

(*a*) 1°K ☐
(*b*) 4°K ☐
(*c*) 0.4°K ☐
(*d*) 0.8°K. ☐

Q. 1191. A temperature of 10^{-2} K can be reached by

(*a*) the Claude Heylandt method ☐
(*b*) regenerative cooling ☐
(*c*) adiabatic demagnetisation ☐
(*d*) adiabatic desorption. ☐

Q. 1192. A particle of mass 5 gm makes 35 oscillations in 11 s, with an amplitude of 8 cm. what is its energy ?

(*a*) 0.0064 joule ☐
(*b*) 0.64 joule ☐
(*c*) 0.08 joule ☐
(*d*) 0.8 joule. ☐

Q. 1193. The unit of angular frequency in S.H.M. is

(*a*) Hertz ☐
(*b*) cycle/second ☐
(*c*) radians/second ☐
(*d*) degrees/second. ☐

Q. 1194. A S.H.M. is given by $y = A \cos (wt + \delta)$. The phase of the motion is given by

(*a*) $\cos (wt + \delta)$ ☐
(*b*) $(wt + \delta)$ ☐

(c) wt
(d) δ.

Q. 1195. Taking the direction to the right as the positive direction of the x-axis, the function $y = f(x + vt)$ represents
(a) a stationary wave
(b) a wave travelling to the left
(c) a wave travelling to the right
(d) a simple harmonic motion.

Q. 1196. A sine wave travelling to the right is represented by the equation $y = ym \sin (kz - wt)$
The velocity of the wave is given by
(a) $\frac{x}{t}$
(b) $\frac{k}{\omega}$
(c) wt
(d) $\frac{\omega}{k}$.

Q. 1197. A musical instrument of frequency 210 Hz is sending out plane waves of amplitude 1/660 of a cm. What is the energy of radiation per cm^3 given that the density of air = 0.00129 per cm^3
(a) 0.00129 $ergs/cm^3$
(b) 0.00065 $ergs/cm^3$
(c) 0.00258 $ergs/cm^3$
(d) 0.210 $ergs/cm^3$.

Q. 1198. The following equation represents a damped S.H.M.

$$m\frac{d^2x}{dt^2} + b\frac{dx}{dt} + kx = 0$$

Pick out the wrong statement:
(a) The frequency $= \frac{1}{2\pi}\ \frac{k}{m} - \left(\frac{b}{2m}\right)^2$
(b) The damping factor is $e^{-bt/2m}$
(c) If damping is small, the frequency is $\frac{1}{2\pi}\ \frac{k}{m}$

(*d*) The damping force is $-b\frac{dx}{dt}$ where *b* is a variable. ☐

Q. 1199. In the case of forced oscillations, the frequency of oscillation is

(*a*) the natural undamped frequency ☐
(*b*) the natural damped frequency ☐
(*c*) the frequency of the external periodic force ☐
(*d*) some other frequency. ☐

Q. 1200. If hydrogen gas is bombarded at room temperature with electrons possessing different energies, the wavelength of radiations first emitted from the gas will correspond to

(*a*) Lyman series ☐
(*b*) shortest wavelength of Lyman series ☐
(*c*) longest wavelength of Lyman series ☐
(*d*) all series will appear together. ☐

Answers

1. (*a*)
2. (*c*)
3. (*a*)
4. (*d*)
5. (*d*)
6. (*c*)
7. (*d*)
8. (*d*)
9. (*d*)
10. (*c*)
11. (*b*)
12. (*d*)
13. (*b*)
14. (*c*)
15. (*d*)
16. (*c*)
17. (*d*)
18. (*d*)
19. (*d*)
20. (*b*)
21. (*a*)
22. (*c*)
23. (*d*)
24. (*b*)
25. (*c*)
26. (*a*)
27. (*d*)
28. (*d*)
29. (*b*)
30. (*b*)
31. (*c*)
32. (*c*)
33. (*c*)
34. (*c*)
35. (*a*)
36. (*b*)
37. (*a*)
38. (*d*)
39. (*a*)
40. (*b*)
41. (*a*)
42. (*c*)
43. (*b*)
44. (*a*)
45. (*d*)
46. (*c*)
47. (*a*)
48. (*b*)
49. (*d*)
50. (*c*)
51. (*a*)
52. (*b*)
53. (*a*)
54. (*a*)
55. (*b*)
56. (*a*)
57. (*a*)
58. (*c*)
59. (*c*)
60. (*d*)
61. (*c*)
62. (*d*)
63. (*a*)
64. (*d*)
65. (*d*)
66. (*b*)
67. (*c*)
68. (*b*)
69. (*c*)
70. (*c*)
71. (*c*)
72. (*d*)
73. (*c*)
74. (*a*)
75. (*c*)
76. (*c*)
77. (*b*)
78. (*c*)
79. (*c*)
80. (*b*)
81. (*a*)
82. (*d*)
83. (*a*)
84. (*b*)
85. (*b*)
86. (*d*)
87. (*a*)
88. (*d*)
89. (*d*)
90. (*c*)
91. (*b*)
92. (*c*)
93. (*d*)
94. (*d*)
95. (*c*)
96. (*a*)
97. (*a*)
98. (*b*)
99. (*c*)
100. (*c*)

101. (*c*)	131. (*a*)	161. (*d*)	191. (*c*)
102. (*b*)	132. (*d*)	162. (*b*)	192. (*c*)
103. (*d*)	133. (*a*)	163. (*d*)	193. (*b*)
104. (*a*)	134. (*d*)	164. (*b*)	194. (*b*)
105. (*b*)	135. (*d*)	165. (*b*)	195. (*b*)
106. (*d*)	136. (*b*)	166. (*b*)	196. (*c*)
107. (*c*)	137. (*c*)	167. (*a*)	197. (*b*)
108. (*a*)	138. (*c*)	168. (*c*)	198. (*b*)
109. (*c*)	139. (*c*)	169. (*c*)	199. (*a*)
110. (*a*)	140. (*b*)	170. (*a*)	200. (*a*)
111. (*d*)	141. (*c*)	171. (*c*)	201. (*a*)
112. (*a*)	142. (*c*)	172. (*d*)	202. (*a*)
113. (*c*)	143. (*d*)	173. (*b*)	203. (*b*)
114. (*d*)	144. (*b*)	174. (*a*)	204. (*b*)
115. (*d*)	145. (*b*)	175. (*a*)	205. (*b*)
116. (*a*)	146. (*c*)	176. (*a*)	206. (*c*)
117. (*b*)	147. (*a*)	177. (*b*)	207. (*a*)
118. (*c*)	148. (*a*)	178. (*a*)	208. (*a*)
119. (*c*)	149. (*a*)	179. (*a*)	209. (*d*)
120. (*a*)	150. (*b*)	180. (*a*)	210. (*d*)
121. (*c*)	151. (*a*)	181. (*b*)	211. (*d*)
122. (*c*)	152. (*b*)	182. (*a*)	212. (*d*)
123. (*d*)	153. (*d*)	183. (*c*)	213. (*a*)
124. (*a*)	154. (*c*)	184. (*d*)	214. (*c*)
125. (*c*)	155. (*b*)	185. (*b*)	215. (*a*)
126. (*d*)	156. (*b*)	186. (*d*)	216. (*b*)
127. (*b*)	157. (*c*)	187. (*b*)	217. (*a*)
128. (*a*)	158. (*d*)	188. (*b*)	218. (*b*)
129. (*a*)	159. (*c*)	189. (*b*)	219. (*c*)
130. (*b*)	160. (*d*)	190. (*b*)	220. (*c*)

221. (*d*)
222. (*d*)
223. (*b*)
224. (*c*)
225. (*c*)
226. (*c*)
227. (*b*)
228. (*c*)
229. (*a*)
230. (*b*)
231. (*a*)
232. (*a*)
233. (*d*)
234. (*c*)
235. (*c*)
236. (*c*)
237. (*b*)
238. (*b*)
239. (*c*)
240. (*c*)
241. (*a*)
242. (*b*)
243. (*b*)
244. (*b*)
245. (*a*)
246. (*a*)
247. (*a*)
248. (*a*)
249. (*c*)
250. (*c*)
251. (*d*)
252. (*c*)
253. (*c*)
254. (*b*)
255. (*d*)
256. (*b*)
257. (*c*)
258. (*b*)
259. (*b*)
260. (*b*)
261. (*c*)
262. (*a*)
263. (*d*)
264. (*c*)
265. (*c*)
266. (*c*)
267. (*d*)
268. (*c*)
269. (*d*)
270. (*b*)
271. (*d*)
272. (*b*)
273. (*b*)
274. (*b*)
275. (*c*)
276. (*d*)
277. (*c*)
278. (*c*)
279. (*d*)
280. (*c*)
281. (*d*)
282. (*c*)
283. (*a*)
284. (*b*)
285. (*c*)
286. (*c*)
287. (*b*)
288. (*c*)
289. (*d*)
290. (*c*)
291. (*c*)
292. (*d*)
293. (*b*)
294. (*c*)
295. (*d*)
296. (*c*)
297. (*d*)
298. (*c*)
299. (*b*)
300. (*a*)
301. (*b*)
302. (*d*)
303. (*c*)
304. (*a*)
305. (*c*)
306. (*b*)
307. (*a*)
308. (*b*)
309. (*a*)
310. (*b*)
311. (*a*)
312. (*a*)
313. (*c*)
314. (*a*)
315. (*a*)
316. (*d*)
317. (*d*)
318. (*a*)
319. (*d*)
320. (*c*)
321. (*b*)
322. (*b*)
323. (*a*)
324. (*c*)
325. (*d*)
326. (*c*)
327. (*a*)
328. (*a*)
329. (*d*)
330. (*d*)
331. (*b*)
332. (*a*)
333. (*d*)
334. (*b*)
335. (*a*)
336. (*b*)
337. (*b*)
338. (*a*)
339. (*c*)
340. (*d*)

341.	(*a*)	371.	(*d*)	401.	(*d*)	431.	(*b*)
342.	(*a*)	372.	(*b*)	402.	(*d*)	432.	(*b*)
343.	(*c*)	373.	(*c*)	403.	(*d*)	433.	(*c*)
344.	(*b*)	374.	(*c*)	404.	(*c*)	434.	(*a*)
345.	(*a*)	375.	(*d*)	405.	(*c*)	435.	(*b*)
346.	(*a*)	376.	(*b*)	406.	(*d*)	436.	(*d*)
347.	(*c*)	377.	(*d*)	407.	(*b*)	437.	(*a*)
348.	(*b*)	378.	(*a*)	408.	(*a*)	438.	(*d*)
349.	(*c*)	379.	(*c*)	409.	(*c*)	439.	(*c*)
350.	(*b*)	380.	(*c*)	410.	(*a*)	440.	(*b*)
351.	(*a*)	381.	(*b*)	411.	(*c*)	441.	(*c*)
352.	(*a*)	382.	(*d*)	412.	(*c*)	442.	(*c*)
353.	(*c*)	383.	(*b*)	413.	(*c*)	443.	(*b*)
354.	(*a*)	384.	(*b*)	414.	(*c*)	444.	(*c*)
355.	(*a*)	385.	(*b*)	415.	(*b*)	445.	(*c*)
356.	(*d*)	386.	(*b*)	416.	(*c*)	446.	(*a*)
357.	(*c*)	387.	(*d*)	417.	(*a*)	447.	(*c*)
358.	(*a*)	388.	(*a*)	418.	(*b*)	448.	(*c*)
359.	(*b*)	389.	(*a*)	419.	(*b*)	449.	(*b*)
360.	(*a*)	390.	(*a*)	420.	(*b*)	450.	(*d*)
361.	(*d*)	391.	(*d*)	421.	(*c*)	451.	(*d*)
362.	(*a*)	392.	(*b*)	422.	(*b*)	452.	(*a*)
363.	(*b*)	393.	(*a*)	423.	(*c*)	453.	(*a*)
364.	(*a*)	394.	(*c*)	424.	(*d*)	454.	(*c*)
365.	(*d*)	395.	(*d*)	425.	(*a*)	455.	(*b*)
366.	(*d*)	396.	(*d*)	426.	(*c*)	456.	(*b*)
367.	(*a*)	397.	(*c*)	427.	(*d*)	457.	(*d*)
368.	(*c*)	398.	(*b*)	428.	(*a*)	458.	(*a*)
369.	(*c*)	399.	(*c*)	429.	(*d*)	459.	(*b*)
370.	(*c*)	400.	(*b*)	430.	(*c*)	460.	(*a*)

461. (*a*)
462. (*b*)
463. (*d*)
464. (*c*)
465. (*c*)
466. (*b*)
467. (*b*)
468. (*c*)
469. (*b*)
470. (*c*)
471. (*d*)
472. (*b*)
473. (*d*)
474. (*d*)
475. (*d*)
476. (*b*)
477. (*c*)
478. (*d*)
479. (*c*)
480. (*c*)
481. (*c*)
482. (*b*)
483. (*b*)
484. (*a*)
485. (*c*)
486. (*a*)
487. (*c*)
488. (*d*)
489. (*d*)
490. (*b*)
491. (*d*)
492. (*b*)
493. (*d*)
494. (*a*)
495. (*b*)
496. (*b*)
497. (*b*)
498. (*a*)
499. (*c*)
500. (*c*)
501. (*d*)
502. (*a*)
503. (*c*)
504. (*b*)
505. (*b*)
506. (*a*)
507. (*b*)
508. (*c*)
509. (*a*)
510. (*b*)
511. (*b*)
512. (*b*)
513. (*b*)
514. (*c*)
515. (*a*)
516. (*b*)
517. (*d*)
518. (*a*)
519. (*b*)
520. (*b*)
521. (*a*)
522. (*c*)
523. (*a*)
524. (*a*)
525. (*c*)
526. (*c*)
527. (*c*)
528. (*a*)
529. (*a*)
530. (*c*)
531. (*b*)
532. (*b*)
533. (*c*)
534. (*b*)
535. (*a*)
536. (*a*)
537. (*d*)
538. (*b*)
539. (*a*)
540. (*c*)
541. (*a*)
542. (*c*)
543. (*c*)
544. (*b*)
545. (*b*)
546. (*c*)
547. (*c*)
548. (*a*)
549. (*c*)
550. (*d*)
551. (*c*)
552. (*d*)
553. (*a*)
554. (*a*)
555. (*b*)
556. (*a*)
557. (*c*)
558. (*b*)
559. (*a*)
560. (*c*)
561. (*a*)
562. (*a*)
563. (*a*)
564. (*c*)
565. (*b*)
566. (*c*)
567. (*c*)
568. (*d*)
569. (*a*)
570. (*c*)
571. (*d*)
572. (*c*)
573. (*a*)
574. (*b*)
575. (*c*)
576. (*b*)
577. (*b*)
578. (*c*)
579. (*a*)
580. (*d*)

581. (*a*)
582. (*c*)
583. (*d*)
584. (*b*)
585. (*b*)
586. (*d*)
587. (*c*)
588. (*d*)
589. (*b*)
590. (*c*)
591. (*b*)
592. (*a*)
593. (*d*)
594. (*c*)
595. (*b*)
596. (*b*)
597. (*b*)
598. (*b*)
599. (*c*)
600. (*b*)
601. (*c*)
602. (*a*)
603. (*a*)
604. (*a*)
605. (*a*)
606. (*c*)
607. (*b*)
608. (*c*)
609. (*d*)
610. (*c*)
611. (*c*)
612. (*a*)
613. (*a*)
614. (*b*)
615. (*a*)
616. (*c*)
617. (*c*)
618. (*b*)
619. (*b*)
620. (*d*)
621. (*b*)
622. (*b*)
623. (*a*)
624. (*a*)
625. (*b*)
626. (*c*)
627. (*a*)
628. (*a*)
629. (*a*)
630. (*b*)
631. (*a*)
632. (*c*)
633. (*b*)
634. (*d*)
635. (*a*)
636. (*c*)
637. (*b*)
638. (*c*)
639. (*b*)
640. (*a*)
641. (*a*)
642. (*d*)
643. (*b*)
644. (*c*)
645. (*d*)
646. (*d*)
647. (*c*)
648. (*c*)
649. (*b*)
650. (*b*)
651. (*c*)
652. (*b*)
653. (*b*)
654. (*d*)
655. (*a*)
656. (*a*)
657. (*d*)
658. (*c*)
659. (*a*)
660. (*b*)
661. (*a*)
662. (*b*)
663. (*a*)
664. (*b*)
665. (*b*)
666. (*a*)
667. (*d*)
668. (*b*)
669. (*d*)
670. (*d*)
671. (*a*)
672. (*c*)
673. (*d*)
674. (*d*)
675. (*d*)
676. (*c*)
677. (*b*)
678. (*d*)
679. (*c*)
680. (*b*)
681. (*d*)
682. (*b*)
683. (*c*)
684. (*a*)
685. (*b*)
686. (*c*)
687. (*b*)
688. (*d*)
689. (*b*)
690. (*d*)
691. (*b*)
692. (*b*)
693. (*c*)
694. (*c*)
695. (*d*)
696. (*a*)
697. (*b*)
698. (*b*)
699. (*d*)
700. (*c*)

701. (*c*)	731. (*b*)	761. (*c*)	791. (*b*)
702. (*b*)	732. (*d*)	762. (*a*)	792. (*a*)
703. (*b*)	733. (*b*)	763. (*d*)	793. (*b*)
704. (*d*)	734. (*b*)	764. (*a*)	794. (*b*)
705. (*b*)	735. (*a*)	765. (*b*)	795. (*d*)
706. (*a*)	736. (*c*)	766. (*d*)	796. (*d*)
707. (*c*)	737. (*a*)	767. (*b*)	797. (*c*)
708. (*b*)	738. (*b*)	768. (*b*)	798. (*d*)
709. (*c*)	739. (*c*)	769. (*c*)	799. (*c*)
710. (*b*)	740. (*a*)	770. (*c*)	800. (*c*)
711. (*b*)	741. (*d*)	771. (*d*)	801. (*c*)
712. (*b*)	742. (*c*)	772. (*a*)	802. (*b*)
713. (*c*)	743. (*b*)	773. (*b*)	803. (*a*)
714. (*c*)	744. (*c*)	774. (*b*)	804. (*b*)
715. (*c*)	745. (*c*)	775. (*b*)	805. (*a*)
716. (*a*)	746. (*c*)	776. (*a*)	806. (*d*)
717. (*b*)	747. (*b*)	777. (*b*)	807. (*a*)
718. (*b*)	748. (*b*)	778. (*b*)	808. (*b*)
719. (*b*)	749. (*b*)	779. (*b*)	809. (*d*)
720. (*c*)	750. (*a*)	780. (*a*)	810. (*d*)
721. (*d*)	751. (*c*)	781. (*b*)	811. (*c*)
722. (*c*)	752. (*b*)	782. (*c*)	812. (*d*)
723. (*b*)	753. (*a*)	783. (*a*)	813. (*c*)
724. (*c*)	754. (*c*)	784. (*b*)	814. (*a*)
725. (*b*)	755. (*b*)	785. (*b*)	815. (*b*)
726. (*c*)	756. (*c*)	786. (*a*)	816. (*d*)
727. (*d*)	757. (*a*)	787. (*b*)	817. (*a*)
728. (*a*)	758. (*c*)	788. (*b*)	818. (*d*)
729. (*b*)	759. (*c*)	789. (*b*)	819. (*b*)
730. (*b*)	760. (*b*)	790. (*d*)	820. (*c*)

821. (*b*)
822. (*d*)
823. (*d*)
824. (*d*)
825. (*d*)
826. (*d*)
827. (*b*)
828. (*d*)
829. (*a*)
830. (*a*)
831. (*b*)
832. (*c*)
833. (*b*)
834. (*d*)
835. (*a*)
836. (*c*)
837. (*a*)
838. (*a*)
839. (*b*)
840. (*c*)
841. (*b*)
842. (*b*)
843. (*c*)
844. (*a*)
845. (*a*)
846. (*d*)
847. (*d*)
848. (*c*)
849. (*b*)
850. (*b*)
851. (*d*)
852. (*a*)
853. (*a*)
854. (*d*)
855. (*a*)
856. (*d*)
857. (*d*)
858. (*b*)
859. (*c*)
860. (*b*)
861. (*c*)
862. (*a*)
863. (*a*)
864. (*c*)
865. (*c*)
866. (*d*)
867. (*b*)
868. (*c*)
869. (*c*)
870. (*c*)
871. (*b*)
872. (*d*)
873. (*c*)
874. (*a*)
875. (*a*)
876. (*b*)
877. (*b*)
878. (*c*)
879. (*b*)
880. (*c*)
881. (*c*)
882. (*b*)
883. (*b*)
884. (*c*)
885. (*d*)
886. (*d*)
887. (*a*)
888. (*d*)
889. (*b*)
890. (*b*)
891. (*b*)
892. (*d*)
893. (*c*)
894. (*a*)
895. (*d*)
896. (*a*)
897. (*a*)
898. (*d*)
899. (*c*)
900. (*c*)
901. (*c*)
902. (*d*)
903. (*b*)
904. (*a*)
905. (*c*)
906. (*d*)
907. (*b*)
908. (*a*)
909. (*c*)
910. (*c*)
911. (*a*)
912. (*b*)
913. (*b*)
914. (*b*)
915. (*c*)
916. (*b*)
917. (*a*)
918. (*c*)
919. (*b*)
920. (*d*)
921. (*b*)
922. (*c*)
923. (*c*)
924. (*a*)
925. (*a*)
926. (*b*)
927. (*d*)
928. (*c*)
929. (*a*)
930. (*b*)
931. (*b*)
932. (*b*)
933. (*b*)
934. (*c*)
935. (*b*)
936. (*b*)
937. (*c*)
938. (*c*)
939. (*c*)
940. (*a*)

941. (*b*)	971. (*c*)	1001. (*d*)	1031. (*c*)
942. (*c*)	972. (*c*)	1002. (*b*)	1032. (*c*)
943. (*c*)	973. (*b*)	1003. (*d*)	1033. (*d*)
944. (*d*)	974. (*b*)	1004. (*b*)	1034. (*b*)
945. (*a*)	975. (*b*)	1005. (*c*)	1035. (*a*)
946. (*b*)	976. (*a*)	1006. (*d*)	1036. (*c*)
947. (*b*)	977. (*d*)	1007. (*a*)	1037. (*a*)
948. (*c*)	978. (*d*)	1008. (*a*)	1038. (*d*)
949. (*a*)	979. (*a*)	1009. (*c*)	1039. (*d*)
950. (*c*)	980. (*a*)	1010. (*c*)	1040. (*b*)
951. (*a*)	981. (*a*)	1011. (*c*)	1041. (*b*)
952. (*b*)	982. (*c*)	1012. (*c*)	1042. (*a*)
953. (*c*)	983. (*d*)	1013. (*c*)	1043. (*c*)
954. (*a*)	984. (*b*)	1014. (*c*)	1044. (*c*)
955. (*a*)	985. (*b*)	1015. (*a*)	1045. (*c*)
956. (*d*)	986. (*a*)	1016. (*d*)	1046. (*b*)
957. (*d*)	987. (*a*)	1017. (*b*)	1047. (*b*)
958. (*a*)	988. (*c*)	1018. (*b*)	1048. (*a*)
959. (*a*)	989. (*c*)	1019. (*c*)	1049. (*b*)
960. (*d*)	990. (*c*)	1020. (*b*)	1050. (*c*)
961. (*b*)	991. (*d*)	1021. (*a*)	1051. (*c*)
962. (*a*)	992. (*a*)	1022. (*a*)	1052. (*c*)
963. (*c*)	993. (*d*)	1023. (*d*)	1053. (*b*)
964. (*a*)	994. (*c*)	1024. (*b*)	1054. (*c*)
965. (*c*)	995. (*b*)	1025. (*c*)	1055. (*c*)
966. (*b*)	996. (*d*)	1026. (*d*)	1056. (*b*)
967. (*b*)	997. (*d*)	1027. (*c*)	1057. (*a*)
968. (*a*)	998. (*b*)	1028. (*b*)	1058. (*a*)
969. (*c*)	999. (*a*)	1029. (*b*)	1059. (*d*)
970. (*b*)	1000. (*c*)	1030. (*d*)	1060. (*b*)

1061. (*a*)
1062. (*a*)
1063. (*d*)
1064. (*c*)
1065. (*b*)
1066. (*c*)
1067. (*b*)
1068. (*d*)
1069. (*c*)
1070. (*b*)
1071. (*b*)
1072. (*b*)
1073. (*a*)
1074. (*a*)
1075. (*a*)
1076. (*a*)
1077. (*d*)
1078. (*d*)
1079. (*b*)
1080. (*d*)
1081. (*a*)
1082. (*c*)
1083. (*b*)
1084. (*c*)
1085. (*d*)
1086. (*c*)
1087. (*d*)
1088. (*d*)
1089. (*d*)
1090. (*d*)
1091. (*b*)
1092. (*b*)
1093. (*a*)
1094. (*b*)
1095. (*c*)
1096. (*d*)
1097. (*a*)
1098. (*c*)
1099. (*c*)
1100. (*a*)
1101. (*d*)
1102. (*d*)
1103. (*d*)
1104. (*b*)
1105. (*c*)
1106. (*c*)
1107. (*b*)
1108. (*b*)
1109. (*d*)
1110. (*d*)
1111. (*c*)
1112. (*c*)
1113. (*b*)
1114. (*d*)
1115. (*b*)
1116. (*d*)
1117. (*a*)
1118. (*b*)
1119. (*b*)
1120. (*b*)
1121. (*d*)
1122. (*c*)
1123. (*b*)
1124. (*d*)
1125. (*a*)
1126. (*d*)
1127. (*b*)
1128. (*a*)
1129. (*b*)
1130. (*c*)
1131. (*d*)
1132. (*c*)
1133. (*b*)
1134. (*a*)
1135. (*c*)
1136. (*a*)
1137. (*c*)
1138. (*c*)
1139. (*d*)
1140. (*d*)
1141. (*a*)
1142. (*b*)
1143. (*b*)
1144. (*d*)
1145. (*a*)
1146. (*b*)
1147. (*a*)
1148. (*c*)
1149. (*b*)
1150. (*b*)
1151. (*a*)
1152. (*c*)
1153. (*d*)
1154. (*c*)
1155. (*a*)
1156. (*b*)
1157. (*a*)
1158. (*d*)
1159. (*a*)
1160. (*b*)
1161. (*b*)
1162. (*b*)
1163. (*c*)
1164. (*b*)
1165. (*c*)
1166. (*d*)
1167. (*c*)
1168. (*c*)
1169. (*b*)
1170. (*a*)
1171. (*a*)
1172. (*a*)
1173. (*d*)
1174. (*a*)
1175. (*a*)
1176. (*a*)
1177. (*c*)
1178. (*a*)
1179. (*a*)
1180. (*c*)

1181. (*c*)	1186. (*c*)	1191. (*c*)	1196. (*d*)
1182. (*a*)	1187. (*d*)	1192. (*a*)	1197. (*b*)
1183. (*d*)	1188. (*b*)	1193. (*c*)	1198. (*d*)
1184. (*d*)	1189. (*c*)	1194. (*b*)	1199. (*c*)
1185. (*b*)	1190. (*c*)	1195. (*b*)	1200. (*c*)